GOVERNORS STATE UNIVERSITY LIBRARY

W9-BFU-668

3 1611 00233 2523

A BRIEF HISTORY
OF THE MIND

**GOVERNORS STATE UNIVERSITY
LIBRARY**

DEMCO

Books by William H. Calvin

A Brain for All Seasons
Lingua ex Machina*
The Cerebral Code
How Brains Think
Conversations with Neil's Brain**
How the Shaman Stole the Moon
The Ascent of Mind
The Cerebral Symphony
The River That Flows Uphill
The Throwing Madonna
Inside the Brain**

*with Derek Bickerton
**with George A. Ojemann

A BRIEF

HISTORY

OF THE

MIND

From Apes to Intellect and Beyond

GOVERNORS STATE UNIVERSITY
UNIVERSITY PARK
IL 60466

William H. Calvin

OXFORD
UNIVERSITY PRESS

OXFORD
UNIVERSITY PRESS

Oxford University Press, Inc., publishes works that
further Oxford University's objective of excellence
in research, scholarship, and education.

Oxford   New York
Auckland   Cape Town   Dar es Salaam   Hong Kong   Karachi
Kuala Lumpur   Madrid   Melbourne   Mexico City   Nairobi
New Delhi   Shanghai   Taipei   Toronto

With offices in
Argentina   Austria   Brazil   Chile   Czech Republic   France   Greece
Guatemala   Hungary   Italy   Japan   Poland   Portugal   Singapore
South Korea   Switzerland   Thailand   Turkey   Ukraine   Vietnam

Copyright © 2004 by William H. Calvin

First published by Oxford University Press, Inc., 2004
198 Madison Avenue, New York, New York 10016
www.oup.com

First issued as an Oxford University Press paperback, 2005
ISBN-13: 978-0-19-518248-4   ISBN-10: 0-19-518248-0

Oxford is a registered trademark of Oxford University Press

All rights reserved. No part of this publication may be reproduced,
stored in a retrieval system, or transmitted, in any form or by any means,
electronic, mechanical, photocopying, recording, or otherwise, without
the prior permission of Oxford University Press.

Corrections and web links for this book may be found at http://WilliamCalvin.com/BHM/.

The Library of Congress has catalogued the cloth edition as follows:
Calvin, William H., 1939–
A brief history of the mind : from apes
to intellect and beyond / William H. Calvin
Includes bibliographical references and index.
ISBN-13: 978-0-19-515907-3   ISBN-10: 0-19-515907-1
1. Brain—Evolution.
2. Cognitive neuroscience.
3. Evolutionary psychology.
4. Human evolution.
I. Title.
QP360.5.C348   2004
153—dc21   2003053093

2   4   6   8   9   7   5   3   1
Printed in the United States of America
on acid-free paper

QP
360.5
.C348
2004

Contents

"Can you tell the story of the world in an evening around the campfire, the way an old-fashioned shaman used to do?" The history of the mind is surprisingly brief. Instead of starting with a big bang, I lead up to one—the "Mind's Big Bang"—and then look beyond, to mind's next advances.

The way we were, 7 million years ago?

Chimps may not be as sociable with humans as a dog that thinks you are its pack leader, or a cat that mistakes you for its mother, but chimp-to-chimp they clearly have a substantial fraction of instinctive human social behavior. They even play blindman's buff. Yet they don't plan ahead very much.

In the woodland between forest and savanna

The dark woods are not where we want to be. We prefer fewer trees, along with a view of some water and grass—which is why waterfront property is now so expensive. Our ancestors were likely digging up veggies, but not making sharp tools. Did the bipedal apes stand upright for the view, to carry the baby, or to avoid taking the midday "heat hit" on the broad back?

In Africa, there was a spinoff with a bigger brain. A new species usually starts out as a small, isolated population. Imagine, say, the big company's branch office in Nairobi losing communication with the parent and having to manage on its own ideas and resources, to sink or swim as an independent in a worsening climate.

Food preparation likely began, maybe even cooking the savory stew. By 1.7 million years ago, *Homo erectus* had spread out of Africa into the grasslands of Asia and was eating a lot of meat. Accurate throwing is a difficult task for the brain. You can't rely on progress reports as you launch (your nerves are too slow). Without timely feedback, you have to make the perfect plan as you "get set"—and there are a million ways to get it wrong, any one of which will cause dinner to run away. So better short-term planning has an immediate payoff. Perhaps that improved their planning for other occasions as well.

When the ice age climate switched into oscillations that were slower and bigger, brain size started growing faster. But why? More demanding hunting techniques? Or ought we to be thinking about the beginnings of protolanguage, the short sentences of modern two-year-olds but without the structuring needed for long sentences?

If the hominids of 400,000 years ago could stage both toolmaking and food preparation, perhaps their life of the mind included other kinds of agendas as well, with more of an eye to the future. Asking whether Neanderthals could speak illuminates some of the changes of the previous million years.

Here we are coming up on the last few minutes of the up-from-the-apes movie, and our vaunted intelligence still hasn't made its first appearance.

Our ancestors were *Homo sapiens* for 100,000 years but, despite the big brain, they were not "behaviorally modern" *Homo sapiens sapiens*. Simple forms of protostructure such as framing and "theory of mind" were likely present, and perhaps the protolanguage like that of modern two-year-olds. Clearly, brain size wasn't sufficient to produce spectacular results. It must have taken something more.

The curb-cut principle and emerging higher intellectual function

In saying "I think I saw him leave to go home," you are nesting three sentences inside a fourth. Other aspects of thought are structured too: multistage planning, games with rules that constrain possible moves, chains of logic, structured music. This structured suite likely aided the giant step up to the modern mind. Did it take another genetic change to become behaviorally modern, or was accumulating culture alone able to trigger the boom by giving babies enough structured stuff to hear so that they softwired their brains earlier? And so excelled as adults?

Was the still-full-of-bugs prototype what spread around the world?

A period between 60,000 to 40,000 years ago looks like the probable time of migration of modern humans into the more exotic parts of Eurasia. And it looks as if they became behaviorally modern in important respects not long before leaving Africa. The lack of time to "debug" the new abilities, before the rough-around-the-edges prototype expanded out of Africa, might be thought of as the first worldwide distribution of crash-prone software.

Higher intellectual function and the search for coherence

We have a fascination with discovering hidden order, with imagining how things hang together. And the problem with creativity is not in putting together novel mixtures—a little confusion may suffice—but in managing the incoherence. Things often don't hang together properly—as in our nighttime dreams, full of people, places, and occasions that don't fit together very well. What sort of on-the-fly process does it take to convert such an incoherent mix into a coherent compound, whether it be an on-target movement program or a novel sentence to speak aloud?

From planting to writing to mind medicine

Once agriculture allowed towns and specialized occupations to develop by 6,000 years ago (the last few seconds of the movie), writing developed from tax accounting about 5,000 years ago. Education now helps us to deal with our fallible minds, to "unlearn" our intuitive but erroneous Aristotelian physics, our intuitive biology of vital essences, and our intuitive notions of engineering that make Darwinian evolution so difficult to comprehend. Medicine now calms the voices and delusions, dampens the obsessions and compulsions, and lifts the depressions. In addition to patching us up, might mind medicine eventually "improve" us?

The moderns somehow got their act together

For fans of how and why questions, a brief summary of the most recent Major Transition in evolution. There are a half-dozen candidates so far for the transition to behaviorally modern *Homo sapiens sapiens*. All may be essential but not sufficient by themselves. The question is not when the last stone is added to the archway but which has a growth curve that becomes steeper and steeper, building on itself. The EvoDevo candidate, those precocious kids softwiring their brains earlier to become more capable adults, could double and redouble the percentage of syntax users in only a few generations.

Inventing new levels of organization on the fly

As an example of four levels, *fleece* is organized into *yarn*, which is woven into *cloth*, which can be arranged into *clothing*. As we advance beyond the one-word level of language after the morning cup of coffee, we begin talking about relationships ("This is bigger than that"). With a second cup, we can advance another level to analogies ("Bigger is better"). Poets have to compare candidate metaphors, however, requiring all manner of superstitious practices in order to shore up their mental house of cards and stabilize a new level.

A combustible mixture of ignorance and power?

Where does mind go from here, its powers extended by science-enhanced education and new tools—but with its slowly evolving gut instincts still firmly anchored to the ice ages? We will likely shift mental gears again, into juggling more concepts simultaneously and making decisions even faster—but the faster you go, the more danger of spinning out of control. Ethics, morals, a sense of "what's right" are possible only because of a human level of ability to speculate about the future and modify our possible actions accordingly. But science increasingly serves as our "headlights," and we are "out driving" them, going faster than we can react effectively.

To Ingrith Deyrup-Olsen, Beatrice Bruteau,
and five of my other informal sounding boards:

the three faux graybeards, and the expatriate cousin,

the neuropsychophilosophers of the psychiatry conference room,

the catalysts of the faculty club,

the surrogates of the think tank,

and those who gather in Jonas' basement boardroom.

The view from the community hearth outside the ceremonial cave?
(Grotte de Font de Gaume, France)

Preface

[History] is concerned not with events but with processes.
Processes are things which do not begin and end but
which turn into one another.

—R. G. COLLINGWOOD, 1939

THERE IS SOMETHING ABOUT a big campfire. Small cooking hearths are very useful but, beginning about 120,000 years ago in South Africa, the archaeologists start finding them supplemented by a bigger hearth. Psychologically, it's very attractive—a community bonfire pulls in people from all round the camp.

Back then, did someone tell origin stories around the campfire? That date, in the middle of the prior warm period in the ice ages, is an enigmatic date, as you'll discover about halfway through my origin story. *Homo sapiens* was around by then—they looked a lot like us, big brains and all—but behaviorally they weren't yet us, the innovative species known as *Homo sapiens sapiens*, a people not doubly wise so much as far more creative. It wasn't until about 50,000 years ago that they were finally doing things that cause us to say, "They must have thought a lot like we do." At that point, they surely appreciated campfire storytelling.

"Can you tell the story of the world in an evening around the camp-fire, the way an old-fashioned shaman used to do?" This was the challenge that the historian David Fromkin took up in writing his short book, *The Way of the World.* It mostly focused, as historians do, on the time scale of civilizations, going back perhaps 6,000 years. There are other admirable short histories which have inspired me, such as Stephen Hawking's *A Brief History of Time,* on the cosmological time scale of the 13 billion years that began with the Big Bang.

My origin story starts at 7 million years ago, so as to cover the time since we emerged from the great apes. To understand the emergence of mind—and particularly the higher aspects of consciousness that so set us apart from the rest of the animal kingdom—we need to understand what the great apes are capable of. And what they don't do. That will help us appreciate what happened in ape-to-human brain evolution since we last shared a common ancestor with the chimpanzees and bonobos of Africa.

It is just in the last 1 percent of that up-from-the-apes period that human creativity and technological capabilities have really blossomed. It's been called "The Mind's Big Bang." In our usual expansive sense of "mind," the history of the mind is surprisingly brief, certainly when compared with the long increase in brain size and the halting march of toolmaking. What came before was not, as we usually assume, a series of increasing approximations to the modern mind. So what set the stage for this creative explosion?

The modern mind of *Homo sapiens sapiens* is so startling, when seen against its evolutionary background, that it is worth the effort to tell the up-from-the-apes story in a space short enough so that all the intermediate stages will linger together in the reader's working memory, reverberating off one another, creating a living contrast that might help illuminate mind's future.

T HERE are many ways to write a book like this, depending on the author's viewpoint. We all tend to deal with the same set of facts, but our intellectual backgrounds and interests differ. Most people writing on the subject were trained in anthropology, linguistics, psychology, or evolutionary biology.

I tend to look at the problem from the standpoint of a neurobiologist, always trying to figure out how nerve cells can analyze the world, make sensible plans for movement, and manage those interneurons that convert thought into action. This is the brain mechanic's time scale of *how*. I was driven to looking into the evolutionary setup for *why* things work the way they presently do. And, since I try to deal with brain circuitry for language and creative plans, I was looking for insights from the comparison of human brains to those of our closest cousins that lack these behaviors. I tend to be impressed by self-organization, emergent properties of neural circuitry, and fast tracks in evolution. For better or worse, this book reflects those issues more than would be found in most books on human evolution. Read widely.

Like most brain researchers, I am inconsistent in using the term 'mind.' Yes, the brain does it all. It is something like the software-hardware distinction—but we are really dealing with the advanced products here, higher intellectual function, and 'mind' is the term that gets across the complexities. This is not a brief history of the brain.

W E tend to see ourselves as the narrator of a life story, always situated at a crossroads between past and future, swimming in speculation. We can construct alternative explanations for how we got where we are, emphasizing one aspect or another as a path. Looking ahead, we imagine various trajectories. We refine our guesses, editing out the nonsense, and achieve a clearer glimpse of our crossroad choices.

Because our less imaginative ancestors couldn't think about the future in much detail, they were trapped in a here-and-now existence. They could anticipate routine happenings (like meals), but not in our extended sense of speculation and worry. No "what if" and "why me?" They were conscious in the sense of choosing between alternative courses of action, but with their unstructured type of mental life, you couldn't narrate a life story or conceive of dying someday. Without creative intelligence, there's no crossroad and no end of the road.

I intend this brief history of the mind to itself be a vista from a crossroads, looking back at simpler versions of mental life, taking stock of what we have now, and then speculating about mind's future. For we are at a crossroads in another sense, that of a frontier where the rules are about to change, where mind shifts gears again.

That's my brief history (you'll have to provide your own campfire). Instead of starting with a big bang, I lead up to one—and then look beyond, to contemplate mind's next advances.

The more we learn about what we are, the more options we will discern about what to try to become. Americans have long honored the "self-made man," but now that we are actually learning enough to be able to remake ourselves into something new, many flinch. Many would apparently rather bumble around with their eyes closed, trusting in tradition, than look around to see what's about to happen. Yes, it is unnerving; yes, it can be scary. After all, there are entirely new mistakes we are now empowered to make for the first time. But it's the beginning of a great new adventure for our knowing species. And it's much more exciting, as well as safer, if we open our eyes.

—DANIEL C. DENNETT, *FREEDOM EVOLVES*, 2003

The Closest Cousins

Above: Orangutan
(Borneo, also Indonesia)

Right: Gorilla (Central Africa)

Below: Bonobos (shown here;
left bank of the Congo) are more
lightly built than the otherwise
similar chimpanzees (not shown;
found in East, Central, and
West Africa).

Some Stage-setting Perspective

IF YOU HAVE TROUBLE with the names, just remember that they nest inside one another: **Animals > mammals > primates > monkeys > apes > hominids > us.**

While **animals** have been around perhaps 800 million years, **mammals** are seen only in the last 200 million years or so. **Primates** evolved from the mammals more than 60 million years ago. Living examples of the early small-brained prosimian forms include tree shrews, lemurs, the slow loris, and the galagos.

Monkeys evolved 40 million years ago from the prosimians. Some of the Old World monkeys lost their tails to become **apes** about 25 million years ago. The ape brain is about twice the size of a monkey brain; apes also have more versatile shoulder joints. The lesser apes, the gibbon and the somewhat larger siamang, are examples of the early apes.

The extant great apes are the orangutan (with whom we shared a common ancestor about 12 million years ago), the gorilla (about 8 to 10 million years ago) and the chimpanzee and bonobo (with whom we shared a common ancestor about 6 or 7 million years ago). They all inhabit forests, though chimps can sometimes be found in the more open woodlands.

Hominids (hominins in new-speak) are all the species between that last common ancestor and us humans. They are upright in posture,

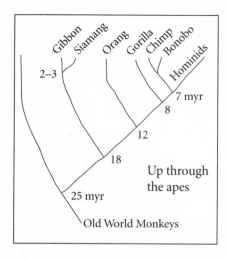

live in the woodlands between forests and grasslands, and have lost the big canine teeth of the apes. But brain size doesn't change much until 2.5 million years ago with the earliest *Homo* species, and that's about when sharp stone toolmaking starts. By *Homo erectus* at 1.8 million years ago, they were eating a lot of meat and were probably inhabiting the grasslands and no longer nesting in trees.

For this brief history of the mind, I will start about 7 million years ago when we shared a common ancestor with the chimp and the bonobo (the misnamed "pygmy chimp" of central Congo), the two great apes with which we have the most in common. The width of the Congo River has kept bonobos isolated for the last several million years from the common chimpanzees, which extend from the East African Rift Valley in Uganda all the way to Senegal in westernmost Africa. Behaviorally, bonobos and chimps have different styles, each of which give us some clues as to what that common ancestor (call it *Pan prior*) might have been thinking, just 7 million years back.

A BRIEF HISTORY
OF THE MIND

The bonobo is overthrowing established notions about where we came from and what our behavior potential is. . . . Even though the bonobo is not our ancestor, but perhaps a rather specialized relative, its female-centered, nonbelligerent society is putting question marks all over the hypothesized evolutionary map of our species. Who could have imagined a close relative of ours in which female alliances intimidate males, sexual behavior is as rich as ours, different groups do not fight but mingle, mothers take on a central role, and the greatest intellectual achievement is not tool use but sensitivity to others?

—FRANS DE WAAL, 1997

1

When Chimpanzees Think

The way we were, 7 million years ago?

WHAT IS IT LIKE, to be a chimpanzee? Are they us, just without language and metaphor? Maybe, as cartoonists suggest, they even talk silently to themselves? (Probably not.)

When we don't understand something like the weather or an animal's mind, we attempt an analogy to something we do know—such as our own familiar mental states or social strategies. Sometimes this works, and sometimes it doesn't. If we don't understand what causes the rain, we may fall back on our knowledge of human social strategy—and assume that a person is in charge of the rain, a person who can be influenced with flattery or gifts or begging. (Sometimes it happens to rain afterwards, and you are persuaded that you were successful.)

And so it is natural enough to suppose that our pets can think. Our pets certainly display emotions, and it is reasonable to assume they have feelings (unexpressed emotions). They certainly have purposeful states of mind—with a single glance, I can tell whether our cat Brin is in her assiduous "mousing mode."

Because dogs are so sociable, we ascribe even more humanlike mental states to them. But we also know that they aren't playing with a full deck of human mental abilities. Despite their considerable spatial abilities—they can catch balls and intercept frisbees in mid-air—most can't untangle their leash from a tree. Nor do most learn from repeated experience to avoid entanglement. Leashes didn't figure in their evolutionary past, and their on-the-fly improvisational abilities just aren't up to the task. Most monkeys have the same problem, though many apes can unwind a tangled leash.

We also feel an affinity to the apes because they romp and play—but then, so do most mammals, at least as juveniles. A bonobo may even play what looks like a solo version of blindman's buff. I've watched Kanzi (the language-reared bonobo in Atlanta), holding a hand over his eyes or draping a blanket over his head—and then showing off his remaining navigation ability.

The bonobos are also sociable in humanlike ways, with much hugging and kissing, reassuring touches, reconciliation after social tension, and doing favors for friends. Unrelated female bonobos build coalitions to control the brotherhood of males. (I once got slapped by Kanzi's sister Panbinisha for moving a big chair; male bonobos grab attention by tumbling heavy objects, and females try to head off such "display behaviors." So I got treated like a bonobo male that was showing off.)

Great apes may not be as sociable with humans as a dog that thinks you are its pack leader, or a cat that mistakes you for its mother, but among themselves our closest relatives clearly have a substantial fraction of instinctive human social behavior. It might have been present in the common ancestor we shared with them, about 7 million years ago.

Y ET there are at least a hundred ways in which the smartest of the apes differ from us. This concentration on making a list of "uniquely human" abilities is a top-down approach, and it attempts to pare away the icing on the cake in order to expose the apelike cake beneath.

There's another complementary approach that might be called "bottom up" as it talks of the perceptual and cognitive abilities common to all the primates, then looks at what emerges with the ape-level mentality, before the hominid icing comes along. Since the rest of the book necessarily has a lot of the uniquely human distinctions, let me start with bottom-up.

The commonalities are extensive. All of the familiar mammals live in pretty much the same world of analyzing sensations and making movements as we do. They recognize familiar objects and they can navigate quite sensibly among them, even taking novel shortcuts to return home (but then, so can ants—with a brain smaller than a pinpoint).

The basic building blocks of behavior, the nerve cells, are pretty much the same in all animals. I've listened in on their chirping electrical conversations in animals ranging from sea slugs to humans, and the basic cellular principles don't change very much. The numbers change, especially in the newer parts of the cerebral cortex where a lot of the associations are done between unlike things. The neocortex is only a few millimeters thick (say, several coins worth) but it is crumpled, with hills and valleys from all the infolding. Were you to flatten out a cortex on a sheet of paper, you'd find that a rat's cortex takes up about the area of a large postage stamp, a monkey's covers a postcard, chimps require a whole sheet of typing paper, and that humans have about four times as much.

Mammalian brains are all laid out similarly, housing the movement command center up front and analyzing sensations toward the rear, with the "where" aspects closer to the top of the head than the "what" conceptual aspects, with the emotional overlay orchestrated from the base of the brain. Furthermore, the circuits and modules become ever more elaborate. Human brains all run on the same hierarchy of chemical and electrical mechanisms but they are compounded into circuits that progress from handling simple categories up to human-only parables.

The first step up from simple perception is lumping things together into "what" categories. Many mammals can create categories of objects, using similarities in color, shape, and texture (but then so can African grey parrots). Some mammals have the rudiments of number sense (usually 1, 2, 3, then skipping to "many"). Though they're not so good at "how" and "why," they can remember "what" is "where" and even "when," as in the case of chimps revisiting distant fruit trees when they are due to ripen.

If you live in larger social groups, remembering "who" may profit from a larger brain. And if you are into sharing things, recalling "who owes what to whom" is important in shaping advantageous social choices and avoiding the inevitable freeloaders. (Everyone loves a freebie.)

It isn't all rote memory or stimulus-response, either. You sometimes see creative inferences and insightful problem solving, especially in the apes. Chimps have enough mental ability to occasionally engage in deception: they can lie because they have some sense that others don't know what they know. That puts them in a new league. *Chimpanzee Politics* and *Machiavellian Intelligence* are serious academic titles about our closest cousins, not publisher's hyperbole.

Even though they pay attention to social happenings in the manner of other primate societies, chimps and bonobos don't augment this with gossip (and more than half of human discourse is catching up on who did what to whom). Chimps throw sticks and stones in an effort to intimidate but rarely as a hunting technique. They are never seen practicing their technique to improve their accuracy or versatility. Nor do the apes exhibit much in the way of shared attention, nothing like the way in which a child directs an adult's attention to a third object. ("Look at that!")

There is also no sustained "paying attention" in special settings, such as our campfire. At least one bonobo accustomed to human ways has learned to feed a small bonfire. But even if the great apes were to master fire starting and had the attention span to tend a campfire for hours, it seems unlikely that they have the psychology to watch a storyteller for very long, even a mime. Too dull, compared to interacting one-on-one, like bored children in church? Or too abstract, involving categories that are simply too many steps removed from here-and-now reality? Or maybe working memory doesn't last long enough?

Of course, just because great apes don't exhibit a behavior in the natural setting doesn't mean that they aren't capable of it, if exposed to the subject by skilled tutors when they are young. If reared in language-rich surroundings, bonobos turn out to be capable of simple forms of word-based communication. They will sometimes point at things, to direct your attention, in a way not seen in the wild. They can understand never-heard-before sentences as complex as "Kanzi, go to the office and bring back the red ball" about as well as a two-and-a-half year old child.

They don't produce such novel sentences themselves, getting stuck (so far, more attempts are in progress) at the stage of two- to three-word sentences and not progressing to longer sentences. A two-and-a-half year old child is, of course, on the verge of blossoming forth into long sentences with syntax. The child is also very acquisitive, picking up nine new words every day in the preschool years and going on to vocabularies of over 50,000 words. The apes have to be laboriously taught new words and seldom learn more than several hundred.

So remember to distinguish between latent capacity and actual practice. The apes are likely capable of many things they don't, in practice, perform—probably because they are not acquisitive of new words and underlying grammar in the manner of human infants. Finding out what is actually impossible for them is a difficult job, which is why you have to treat "uniquely human" claims as provisional.

The archaeologist Steven Mithen, in his *A Prehistory of the Mind*, compares the chimpanzee's termite fishing stick with a human's line and reel, with specialized hooks and weights. The chimp manufactures the tool using the same hand and arm movements as used for other behaviors, while the actions that humans use to carve a spear or craft a bowl are unique gestures. Chimpanzees don't think up new functions for the same tool, the way a human will use a cutting tool to dig dirt out of a small crack. And when a new tool or gesture is invented, other chimpanzees are slow to pick up its functionality. In humans, imitation rapidly spreads the innovation through the population, sometimes improving it via copying errors.

W E tend to assume that bigger brains are better—so that apes, with twice as large a brain, are an improvement over Old World monkeys. Not by some measures: monkeys often outcompete apes and, over the last 10 million years, the number of ape species has been declining while the number of monkey species has been increasing.

Head to head with the chimps of Uganda, monkeys can strip the trees of ripening fruit faster, they can reproduce more frequently, and so forth. Perhaps the ape brain is more versatile in some situations, such as improvising during climate crashes, but—except for eating one occasionally—the smarter apes sure don't dominate the monkeys in the business-as-usual forest settings.

And the apes don't necessarily make use of their bigger brains. The gorilla needs his extra long gut (to extract calories from that 50 pounds a day of low-quality plant food) far more than his big brain. The omnivorous chimps and bonobos can readily switch what they eat, having both a multitalented brain and a more versatile digestive tract. But the other surviving great apes have settled into vegetarian niches that don't require their intelligence. Ditto the marine mammals, which don't need their bigger brains for making a living by filtering plankton or catching smaller fish, for which fish-sized brains seem to suffice. But that doesn't mean that brains downsize accordingly, to save on calories. Backing up isn't easy in evolution.

Did you ever walk into a room and forget why you walked in? That's how dogs spend their lives.

—THE COMEDIAN SUE MURPHY

THINK ahead. The apes do some of it, but how far ahead, and in what detail?

Multistage planning involves much more than what you see in a squirrel preparing for winter by hoarding nuts. That's just a simple instinct that every squirrel is born with, triggered by the days getting shorter and more melatonin being released from the pineal gland every night because darkness lasts longer. It's a wonderful biological example of that lesson from mindless automata, that complicated patterns can arise from the interaction of several simple relationships.

No learning, no planning—a timeless here-and-now mental life ought to suffice for the squirrel, with the melatonin simply providing a lingering bias to what most interests the squirrel during the daytime. Planning usually involves novelty, not something seasonal that all of your ancestors had to do without fail. Getting ready for winter is too important to be left to learning or improvisation.

All animals have a behavioral repertoire. They can focus on one behavior and hold it ready, like a cat about to pounce, or a monkey ready to grab the fruit when the dominant male finally looks the other way— but they are not planning in depth or detail, something we humans are very good at. The great apes have some migrations that look, at first, as if they might qualify as planning. But like most seasonal migrations, the leaders have been over the track for some years in the past as followers. That's learning, not planning. Young chimps learn a minor form of staged food preparation, what it takes to crack tough nuts rather than shatter them. But it takes them six long years of fumbling practice, not a moment of insight followed by a marked improvement in technique. Outside of the half-hour time scale of intentions, a chimp or bonobo doesn't seem to prepare for tomorrow.

It was initially supposed that planning and communication would have big payoffs for organizing a hunt. Chimpanzees (and bonobos, though there is much less data) do hunt small animals, mostly monkeys and bush pigs. They seem to have all the basic group moves, someone covering each of the possible escape routes of a treed monkey.

Creativity is also infrequent; there is no evidence for an ape planning a novel course of action in any depth. (This, too, could be overturned next year; I'm just reporting on the current lack of evidence.) We anticipate our next handhold in climbing a tree, but the really difficult versions are when you have to plan multiple stages of the action in advance, rather than just groping your way along while guessing one stage ahead. The driver who uses grand slalom tactics in freeway traffic, leaving a trail of flashing brake lights in his wake, does not really need higher intellectual function to assist him, only the apelike abilities to swing through the trees, looking ahead to the next handhold. Planning in depth is what I am focusing on here, what you need to imagine several preparatory stages, as in a college course schedule or a new crop rotation.

What chimps don't do gives us some insight into their planning capacities. If chimps could plan ahead, they would be the terror of Africa (and probably extinct by now), but they're not. And you can't just argue that they're peaceful. Chimpanzee groups actually patrol the boundaries of their usual territory, looking for all the world like an army patrol that re-forms into single file, keeps quiet, and stops occasionally to listen carefully before moving on. They may engage in shouting matches with the neighbors, judging group size, but they seldom get into battles. Yet when the chimp patrol finds a lone chimp from the neighboring group, you see what looks like human gang warfare, five-on-one affairs that leave behind a dying chimp, its throat or genitals chewed out, great strips of skin pulled loose. (Bonobo groups are larger and they often mingle peacefully when encountering one another. They have not been seen patrolling their borders.)

With a little foresight added to that aggressiveness, chimps could conduct raids in the middle of the night. A little more and they could make war on whole groups of neighbors using stockpiling of supplies, practiced maneuvers, and coordinated attacks. But they don't.

It may be that, instinctively, humans are less violent than chimps— but our planning abilities certainly amplify what violent tendencies there are, as does our propensity to form up in ad hoc teams spontaneously (truly impressive in emergencies, but formed instead into fan clubs they sometimes happily beat up another such ad hoc group).

No one sees much evidence of logical planning in the chimps, and certainly not the sort of planning (so handy for serious warfare) where two or three novel stages have to be worked out in advance of acting. They probably lack complex thought, as did the bipedal woodland apes of our ancestry. The big question is when this plan-ahead capability arose in hominid evolution.

Learned staging and innovative on-the-fly staging are, perhaps, different things that evolved at different times. Much of my virtual campfire tale is about how slowly we acquired those two types of staging. The hard part turns out to be the innovations. While it is easy to create random variations, it is much more difficult to discover, offline, the combinations that are safe and useful. You need *coherence*, where a lot of things fit together satisfactorily, before acting. That's what is probably not well developed in the apes, or in the woodland ape that walked upright in the late Miocene. (And maybe such coherence competence wasn't even present in modern-looking humans of 100,000 years ago.)

A sense of the future may also have been missing at this stage of human evolution. I don't mean this in any simple sense like squirrels hoarding for winter. Nor in the "what happens next" sense that our cat must experience when she sees two adults heading her direction with a medicine dropper in hand. That she has a pretty good notion of what happens next is just learning. It doesn't mean that she is likely to speculate about losing her teeth in old age. Nor is she capable of reflecting on the fact that losing the last set of opposing molars is how elephants die.

I think that, before structured thought, our ancestors mostly had a here-and-now mental life with little structured interpretation of the past. And probably not much on-the-fly contingent planning. They saw death every day but, without much ability to speculate about the future, they couldn't conceive of their own mortality.

The chimps lack language and symbol, virtually lack true teaching, and [there is no] evidence of the sort of metacognition—awareness of mental process—that is the essence of human culture.

—MELVIN KONNER, 2001

In one sense, the bonobo is a fifth subspecies of chimpanzee (they can all interbreed). But the bonobo (shown here), isolated on the left bank of the Congo from the other chimps for about 2 million years, developed enough differences in anatomy and behavior to be called a separate species, *Pan paniscus*. They are also endangered by forest clearing and hunting (the bushmeat trade follows the new roads). The other chimpanzees, *Pan troglodytes*, and the gorilla are now critically endangered by both hunting and an Ebola epidemic.

Woodland, gallery forest, and savanna in the East African Rift Valley
(balloon view of Maasai Mara, Kenya)

2

Upright Posture but Ape-sized Brains

In the woodland between forest and savanna

T HE DARK WOODS IS not where we want to be. Fairy tales play right into this predisposition of ours, seen even in children without experience of the forest or the savanna. We much prefer a few trees, together with a nice view of some water and grass—which is why waterfront property is now so expensive.

Bonobos would not agree with us. They live in equatorial forest, as do chimps, and they likely find wide-open spaces somewhat threatening, just as we do deep dark forests. There are a few places, such as Senegal in the west of Africa and Malawi to the southeast, where chimps live in woodland settings. In a woodland, the trees are interspersed with open areas. That's also what you find on the fringe of a forest, as a transition between packed trees and grassland. There are, of course, various mixes such as tree savanna and bush savanna, but let me simplify the real estate into forest, woodland, bushland, grassland, and really arid desert where little grows (like the modern Sahara).

Over a period of about 5 million years, our ancestors presumably made the transition from feeling comfortable in the forests to being reasonably comfortable out on the savanna, where *Homo erectus* was making a living about 1.8 million years ago. You can tell something about diet from where the fossils are found—not from how the fossil site is today (often eroded badlands), but from the nearby bones of other species whose habitat is known to be forest, woodland, or grassland. For most of that 5-million-year-transition period, the fauna associated with hominid bones are woodland creatures such as pigs, which root around for underground resources that are characteristic of woodlands, not forests or grasslands.

B IG brains come along later in the game. What distinguishes the early hominid from the great ape tends, in the eyes of the physical anthropologists who argue about this sort of thing, to be two things: the reduction of the canine teeth and the acquisition of upright posture.

The pelvis becomes more bowl shaped. Unlike chimps where the legs go straight down from the hip joint, the hominid knees indent. The hole in the bottom of the skull (where the spinal cord descends into the neck) moves forward, to better balance the head atop a now-vertical spinal column. The neck muscles insert into the skull somewhat differently. Some of those upright characteristics are now seen in hominid fossils as early as 6 to 7 million years back.

That isn't to say that they had bipedal locomotion as we know it now. By *Homo erectus* times, they did, but there may have been 5 million years of transition between the occasional upright locomotion of chimps and an efficient upright stride. Our present bipedal locomotion is in fact more efficient than the ape's quadrupedal style, but I doubt that the transition was all about "progress" as some transitional stages seem inefficient. (And Darwinian adaptations for efficiency have to operate on immediate opportunities, not long-term prospects.)

WHAT started upright posture, and why is it associated with woodland habitats? In the old museum tableaus, early man stood upright to peer over the tall grass and brush, all the better to spot prey and predators. While formulated for the savanna theory (the notion that we descended from the trees and strode out into the grasslands), it still works pretty well for the new intermediate stage in woodland. Woodlands were surely comforting when venturing out of the forests, as there was often a handy tree to climb, both for escaping a lion and for nesting at night.

Scavenging is often mentioned as part of the transition to the more serious hunting in *Homo erectus* times. And woodland is a good place to steal some meat, then run off to climb a tree, so as to avoid conflict with late-arriving lions and hyenas. Running fast is useful, and chimps aren't very good at it when carrying something. They waddle, shifting their weight ponderously, and so they tire easily after a short burst of running.

Another reason for being upright in the woodlands is sunshine. In a shady forest, it is hard to get overheated. But in a woodland, most places get some sunshine for part of the day, thanks to the openings. This means that you cast a shadow, and the size of the shadow at midday says something about how much of a "heat hit" you are getting from direct sunshine. Stand upright in the tropics and a small dark pool is seen near your feet.

Grazing animals cast big shadows but, unlike the bonobo shown here in the unnatural setting of a zoo "savanna" at midday, their brains are adapted to heat stress. If our brains were subjected to the body temperature of an eland at midday, we'd have a seizure. So one way to avoid excess heating is to present a minimal target to the sun. If you stand up straight, your head and shoulders take the hit, and they're a lot smaller than your back.

Overheating can also be combated by sweating. Evaporative cooling works best with minimal body hair; sweat that evaporates from a hair doesn't cool the skin very much, for the same reason that the handle of a pan doesn't transfer very much heat to the hand that holds it. You want the sweat to stay on bare skin, so the heat transfer does some good in cooling the body beneath.

But the loss of body hair, another one of those things that changed sometime between the apes and us, has an important consequence for posture. Transporting infants is generally accomplished in the quadrupedal monkeys and apes by the infant clinging to the mother's hair, so she can get around on all fours. Thus the transition to profuse sweating likely resulted in a mother having to use an arm to hang onto the infant, and rearrange her travel posture accordingly.

Another consequence of sweating so much is having to stay close to drinking water. Some animals have kidneys that are very efficient at keeping water from being lost in the urine, but we're profligate, wasting both water and salt and thus constantly having to seek out such resources. It makes me think that our ancestors were often waterhole predators, in competition with the big cats.

So the early hominid habitat was likely a transition zone between forest and grassland, the place where we adapted to heat stress and learned to eat a different diet. Upright posture was likely a byproduct of such factors. What's surprising to many of us is that the postural rearrangement comes so early, back when the DNA dating suggests we parted company with the ancestral chimps. And millions of years before bigger brains developed.

Teeth tell tales too. Besides upright posture, it is clear that some-thing was also going on with the teeth. You can tell whether it is an ape or a hominid by the size of the canine teeth. Since big primate ca-nines are primarily for fighting, and threats to fight, smaller ones in the hominid line suggest that something was going on that made such ag-gression less important.

Was it monogamy, like the gibbons? Probably not, because the size difference between males and females changed in the wrong direction. At both ends of our 5-million-year-long spectrum from upright apes to *Homo erectus*, one sees males that average about 15 percent larger than females, same as modern humans. In between, however, most upright ape species so far seem to have males that are almost twice as large as fe-males. (Think gorillas.) In the animal world, such sexual dimorphism is usually because the males fight, to exclude one another from access to females, which makes for gorilla-like harems. In that game, bigger is better. With *erectus*, the size difference becomes minor, more like today.

I'm not sure what to make of this puzzle. One possibility is that the hominid species with oversized males are not actually our ancestors, that there is an undiscovered lineage somewhere, perhaps outside the Rift Valley, which evolved without significant changes in sexual di-morphism.

And there's another reason to wonder about that, because the teeth also go back and forth in that 5-million-year span, becoming much larger and then much reduced. That may have to do with what there is to eat in woodlands. The lack of all-day shade in the open woodlands means that the sun can dry out the soil, in a way that forest floor plants don't experience. So woodland plants have a lot of underground stor-age organs for water and building materials. We call them bulbs, tubers, rhizomes, or just "veggies." Even chimps that live near woodlands have been seen digging them up in the dry season, if just for their water. Were the chimps to eat them more regularly, the variants with larger cheek teeth might fare better.

But what with *Homo erectus* at the end of the 5-million-year transition having smaller cheek teeth, we again have a back-and-forth situation. Maybe it is just adaptation tracking the diet, maybe it is another sign that we've missed a more direct hominid lineage to *Homo*, with the australopithecines and such off on the side branches. Only more data, earned in the hot and dusty badlands, are likely to settle the issue.

Freeing up the hands presumably had some important effects, as Darwin speculated, but brain size sure didn't change much for the next few million years. So the thought processes of the bipedal apes may have been no fancier than those of the great apes. There might have been chimplike tool use, but it is difficult to find evidence of making stone tools until the very end of this period, at about 2.6 million years ago. An ape-level mentality might have sufficed for life in the woodlands.

The *Homo* lineage is a spin-off, to use a modern term, of the lineages of the bipedal woodland apes. It occurs about 2.4 million years ago. The bipedal apes keep going, evolving into more heavily built vegetarians, until they die out about a million years ago. It's hard not to think of them as a woodland version of the gorillas, specializing themselves into an evolutionary dead end. That's the usual fate of many species and it tends to be aspects of mind—such as having the omnivore's wide set of food-finding tactics—that can provide the versatility needed to avoid getting trapped.

Emotions prepare all organisms for action, for approaching good things and avoiding bad things. But when we step away from the core emotions such as anger and fear that all animals are likely to share, we find other emotions such as guilt, embarrassment, and shame that depend critically on a sense of self and others. I will argue that these emotions are perhaps uniquely human, and provide us with a moral sense that no animal is likely to attain.

—MARC D. HAUSER, 2000

Hyena carrying away meat, trying to escape the other Serengeti scavengers.

Our ancestors likely tried the same tactic. Once they knew how to create stones with sharp edges, they could quickly amputate a leg at a joint and run with it, leaving the rest to the lions and hyenas. Limb meat is initially quite sterile and uncontaminated by opening up the belly.

3

Triple Startups
about 2.5 Million Years Ago

Flickering climate, toolmaking, and bigger brains

I N AFRICA, THERE WERE a lot of new species about 2.5 million years ago, when three big things happened: stone toolmaking, the ice ages, and a hominid spinoff with a bigger brain.

Climate had been gradually cooling and drying in the habitat of the upright ape. This made the forest patchy where it had been continuous, with forest surviving only in particularly good soil or in the uplands that effectively tickled the passing rain clouds. This subdivision meant that there was more forest edge than ever before, thus creating more woodland and more woodland apes.

Between 3 and 2 million years ago, climate became chaotic. Up north, the winter sea ice reached as far south as France 2.51 million years ago, which is as good a time as any other to say that the ice ages "began."

In Africa, the problem was not ice. It stayed confined to the upper reaches of such places as Mount Kenya and Mount Kilimanjaro, visible on a clear day but not intruding into most hominid lives. At sea level, the air temperature was only a few degrees cooler—more comfortable if anything. The African problem was that less rain fell. At times, the rain forests shrank by 80 percent. Major East African lakes, even immense Lake Victoria, eventually dried up.

Most of what we know about hominid evolution in this period comes from the limestone caverns of South Africa or the old lakeshores of the Rift Valley up north, especially a series of fossil sites ranging from Laetoli, 3 degrees south of the equator, to Hadar at 11 degrees to the north. Early hominid fossils have also been discovered to the north of Lake Chad, a half-continent to the west of the Rift. Forests are unlikely to preserve bones; our best views into hominid evolution occur at places where either sediments from a flooding lake bury a corpse, or where someone gets trapped in a cave and the stalactites drip on their bones, sealing them in stone.

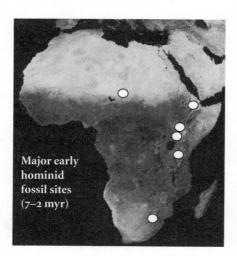

Major early hominid fossil sites (7–2 myr)

We have little choice but to sympathize with the drunk who searched for his lost keys under the streetlight, because the light was better there. We don't know where the lost bones of the earliest hominids are but have to satisfy ourselves with what we can find where preservation is better and, since no one has the money to dig deeply into the earth, in the present-day badlands where the surface erosion exposes the sediments of those ancient lakes.

Looking for ancient tools follows pretty much the same imperatives, prospecting where the ancient lake edges were or the shelters nearby. Much of the dating of these sediment layers comes from the easily dated volcanic layers, which often sandwich the artifact layers of inter-est. The flora and fauna of a period in volcanic East Africa often allow the hominid fossils in the caves of South Africa to be dated, via those associated fauna that spread between East and South Africa.

From such fortuitous circumstances come our occasional glimpses of the state of mind of our ancestors. Could they innovate? Could they postpone advantage (as in delaying consumption, in order to prepare the food better)?

It is thought that diet became less vegetarian, with more high-quality food such as meat, about this time. Some view eating meat as a setup for brain size increase, and it's an interesting line of reasoning. Gorillas have such long guts (and big bellies) because their leafy diet is low in quality, rather like "calorie free" zucchini. So a long gut gives the food more of a chance to give up its calories to the intestinal wall. This also means that the gut needs a lot more blood supply, just to service the long "disas-sembly line." If, so the reasoning goes, you improve the diet with meat (that is readily broken down and absorbed), the variants with shorter gut lengths would still get by. Those variants with both shorter guts and bigger-than-average heads would then have the extra blood flow needed by the bigger brain.

Some marsupials devote less than 1 percent of what their heart pumps to servicing their brains. In the average mammal, it is about 3 percent. In humans, it is close to 16 percent. If something had to give, it was likely the overextended guts. An ancestral diet of low quality food, on this argument, keeps bigger-brained variants "vegetating" while di-gesting a meal.

So the dietary shift into more meat, while not "causing" bigger brains, helps to clear the path for any bigger-brained variants that come along. It doesn't say anything about whether this meat was scav-enged or hunted. Given all the late-arriving lions, hyenas, jackals, and vultures, any hominid lucky enough to kill a grazing animal would shortly face some competition. Defleshing using sharp tools leaves behind "cut marks" on bone, and such cut marks have been found as early as 2.5 million years ago in Ethiopia. They are on the bones of large animals, even hippos—not the small monkeys that chimps and bonobos opportunistically hunt.

Whether hunting or scavenging, it represents a major behavioral change because of dealing with the competition. For these hominids, it wasn't just a manner of avoiding becoming dinner for a lion but a problem of how do you chase them off or avoid them. Tearing a piece loose, and making off with it, is a common strategy. (You need to leave enough behind so that the competitors are more attracted by it than you.) Hyenas can be seen to do this in Ngorongoro Crater, before the lions arrive to take the carcass away from the hyenas that made the kill. (No, that's not backwards; in the nearby Serengeti, it is true that the lions hunt and the hyenas primarily scavenge, but they switch roles in the crater highlands. In general, many scavengers can hunt and many hunters happily scavenge.)

Making meat quickly portable might have been an early application for fractured stones with a sharp edge. Bash one rock into another, and then search for an edge sharp enough to amputate a leg at the knee or hip. (Indeed, just throw a rock against another rock and then search the fragments. But aimed blows are less wasteful of raw materials than the primitive "shatter and search" technique.) Then run with the limb and haul it up the nearest tree, allowing the lions and hyenas to be distracted by the rest of the carcass.

It also allows one to swing the leg like a club, pounding a long bone against a rock outcrop or tree trunk until a spiral fracture develops (such fractures also date back to 2.5 million years ago). Opening up a long bone allows access to the marrow, a highly desirable source of dietary fat, which comes out looking like a long pink sausage. Fat is usually in short supply on the savanna, and a lot more of it is needed for building a bigger brain because the insulation wrapped about nerve cells is fatty.

Stone toolmaking is first seen about 2.6 million years ago in Ethiopia, and the earliest of the bigger-brained *Homo* species have been traced back to 2.4 million years. So first it's toolmaking, then the spinoff. While the dates may subsequently leapfrog one another as new evidence emerges, that is a typical ordering: form follows function. Behavior invents, using the klutzy old anatomy, then anatomy slowly changes to make the new behavior more efficient via the usual Darwinian process.

CONSIDER the plight of the spinoff. Life as a spinoff is not easy. No, I have not shifted my origin story to the gas-log corporate campfire of the present day. I am just beginning a little parable about the hazards of being a new species, isolated from the generous reservoir of culture and genes available from the "parent company."

A new species usually starts out as a small, isolated population. Imagine, say, the big company's branch office in Nairobi losing communication with the parent and having to manage on its own somehow. No more imperatives to conform but also no more resources to draw upon, no buffer against hard times locally. In biology, this is particularly precarious, as the usual swings in population with bad weather may take you down to zero—wipeout, with no hope of recovery. The only thing that saves many small populations is reconnecting with the large parent population. The infusion of immigrants will bring some of the gene variants that were lost simply by chance in the downsizings.

But the immigrant infusion may also reverse "progress," as when Nairobi's independent way of doing things is, upon communications being restored, re-conformed to the home office practice. With all those forms to fill out again and auditors auditing the auditors, the innovative Nairobi office loses whatever edge it had developed during its isolation as an independent.

BRAIN size is carefully regulated in the course of growing up inside the womb. Starting halfway through pregnancy, the trajectory of human brain size rises above that of an ape. The brain becomes relatively bigger, compared to the body. So this is not recapitulation, where you first produce a monkey brain and then tack on some changes in the second act to keep enlarging the brain. No, this is tinkering with things partway through the first act, early in gestation, much the way that the innovation of indoor plumbing eventually changed the way they built new houses, even in relatively early stages.

Is this a new "gene for big brains" that has done the job? That's the usual way we tend to talk about it but, if we mean to be literal rather than metaphorical, it may well be wrong. The regulation of the brain growth trajectory, if it is like other aspects of bodily development, is likely to be done by a committee of genes. Some push, others pull, creating a channel that brain growth usually follows. So instead of a new version of a gene, the big brain change might just be the elimination of one of the genes that held down the growth trajectory.

As more of the chimp genome is sequenced and compared with the human genome, we will be looking for eliminations as well as additions. Or, more likely, we will find the ascendancy of a less efficient gene variant. Many genes come in various versions (called alleles), and that is how faster and slower, bigger and smaller, are usually varied in the body that the genes produce.

For example, the receptors ("locks") on the surface of a brain cell—the ones that sniff out the neurotransmitter molecules ("keys") released from upstream cells—can be made in several different ways. The A_1 allele makes a protein that is 30 percent less efficient than that made by the more common A_2 allele of the D_2 dopamine receptor. This turns down the sensitivity of the downstream cell to dopamine. This less sensitive A_1 allele is found in a quarter of the population, and it may predispose its possessor to certain kinds of addictive behaviors (not just drugs but also gambling and obesity). Perhaps it is merely a matter that these individuals do not satiate as easily, and their higher levels of ingestion thereby set off some chronic changes.

The history of the field suggests that this preliminary explanation will be supplanted soon, but I intend it only as an example of how evolution works by having alternative versions. Say, bluegrass and crabgrass. The environment (how often you water and mow) tends to make some variants more common and others less so (which is what Darwin meant by his term "natural selection").

Let us now suppose that our isolated branch office, instead of dying out, is thriving at the time that lines of communication are restored. Usually immigrants restore things and any idiosyncratic ways of doing business developed by the branch office in the meantime are conformed to the standards of the parent company. But sometimes, the branch office people just don't mix well with those from the parent organization. The locals have become independent (a "new species") and, rather than reintegrating, they go into competition with the "parent company." Size usually wins such competitions but occasionally the independent upstart will expand worldwide, maybe even forcing the parent company out of business.

The economists would likely ascribe the success of the long-lost branch office to an "entrepreneurial culture." But in my parable, it was more a matter of losing a lot of the corporate culture, allowing a reformulation. Gene versions get lost when things downsize, and sometimes that can make a big difference—such as setting brain growth in motion.

Note that an archaeologist sampling the history every few years would likely get the story wrong. What would be seen in most places would be the "sudden" replacement of the old-line company by the newbie. It would look like the old company suddenly transformed itself into the newbie. Only by sampling the history of the long-lost branch office would the gradual changes be seen, of both the isolation and the expansion of the new species. And only a particularly fine sleuthing job would disclose when it was loss, rather than innovation, that played a key role in their improved corporate culture. End of parable.

With all of the droughts in Africa, hominid populations must have often been isolated from one another for considerable periods. Once they developed the skills to make a living in the grasslands, they would have taken advantage of the aftermath of a forest or brush fire, when new grass replaces the burnt woody plants for a few decades. It was a good setup for getting stranded when the grasslands shrank in response to plant succession, all those woody plants starting to return and shrinking the size of the grazing herds. There were many opportunities for a branch office to become independent—or go extinct.

No one really knows yet what toolmaking, a bigger brain, and more chaotic climate have to do with one another, but these relationships give one some food for thought. Was the hominid of 2.5 million years ago still thinking pretty much like an ape, just with woodland overlays?

They probably became much more daring, having to play games with those lions. It became much more important to know what you could get away with. They surely had an overlay of advanced hunting instincts by then, going considerably beyond the instinctive group maneuvers seen in chimps and bonobos when they hunt. And their social instincts had likely changed as well, with more cooperation and sharing.

Even when we think we can back it up to an original quantum fluctuation [as the origin of the universe], we still have to deal with the fact that human beings, who have complexified out of the products of that fluctuation, are beings that concern themselves with issues of meaningfulness and value. We insist on asking "How come?" meaning "What for?" meaning "What ought we to do?"

—BEATRICE BRUTEAU, 2003

Top: Baboons are monkeys that have adapted to savanna foraging.
(Maasai Mara, Kenya)

Bottom: The major savanna resource, however, is grass and large grazing animals.
(Wildebeest herd, Ngorongoro Crater, Tanzania)

4

Homo erectus Ate Well

Adding more meat to the diet fueled the first Out of Africa

WHILE IT IS NATURAL to focus on survival during down-sizings, abilities such as hunting are more likely to be shaped by the opportunities during the aftermath of an abrupt climate change such as a drought. Death on the downside may have more to do with bad luck—being in the wrong place at the wrong time—than it does with not having the right stuff.

Opportunities occur on the upside, the very next year, but only for those with the right stuff. You can only expand into newly burnt territories if you can get by for awhile almost entirely on hunting (a modern example would be the Inuit along the Arctic Circle). So only a fraction of the surviving population can expand; the others stick to the remaining refugia where they can make their living in a more traditional way. The ones that expand are the ones that get the opportunities to become stranded—and occasionally become a new species.

Hunting abilities may provide an optional part of our ancestral diet—there are many studies showing that meat is a minor part of the calories in hunter-gatherer societies—but hunting does get sub-populations into new territories like nothing else, thanks to grass being the low end of the chain of plant succession. This is my major reason for being suspicious of the gathering-is-more-important argument. It assumes efficiency as the main driver, and infers that the major portion of the average diet is more important than the occasional scraps of meat. But dynamics and statics give you different views of the problem, especially when climate changes are so fast that there is a lot of random death followed by a lot of selective opportunity. Having the "right stuff" is, however, quite important in the grassy milieu created by a sudden drought—not just for thriving where others cannot, but for getting into situations where becoming stranded is a common aftermath.

SUSTAINED attention is common in hunting animals. Our cat Brin will spend hours perched atop a fence, watching intently for movements in the grass. She will occasionally dash into the house for food and water, then go right back outside to her "project."

Not much of chimp and bonobo hunting involves such sustained attention; their hunting is more episodic, spontaneous and opportunistic. If they had some of the moves of the big cats, their diet would likely be more than 5 percent meat.

Certain aspects of intelligence don't have much effect unless you have the attention span to go with them. Since great apes are not especially known for such sustained attention, we might infer some changes in that direction by the time of *Homo erectus* when they had adapted to the hunting life. Having both a bigger brain and the hunter's versatile attention span might have opened up new avenues of mental life.

Joint attention (where an individual directs another's attention to something and they jointly contemplate it) is not much seen in the great apes but emerges in early childhood for us moderns. Some researchers would nominate it for the prime mover in the transition to behaviorally modern humans (more later).

Another kind of attention might have developed about then, that needed by the nighttime sentry. Back in open woodland, there were still trees in which to nest at night. While a leopard might pick off an occasional sleeping individual, the scattered nature of the nesting sites meant that the others had enough warning to wake up and run. But out on the savanna, there usually aren't enough trees to go around, making it easy for an entire hominid group to be ambushed, much in the manner that Hadza hunters of Tanzania today ambush groups of baboons that lack trees to nest in.

Occasionally staying awake while others sleep, and having others depend on you to do so—enforced by social sanctions—certainly looks like an advanced social behavior that goes beyond what we share with the chimps and bonobos, yet one that need not wait for language abilities.

The other thing that suggests some social change is the move to a modern amount of sexual dimorphism. Perhaps coalition behaviors were often capable of keeping one big alpha male from excluding the other males from mating opportunities. This loss of the advantage of big male body size might have brought them closer to the modern amount of size difference between males and females.

A FTER the spinoff of the *Homo* lineage, there may have been a number of different hominid species at the same time—exactly what one expects when a new niche is discovered. The meat-eating niche was, for hominids, such a new niche event. For the next half-million years, variants like *Homo rudolfensis* and *Homo habilis* surely tried out their combination of brains and guts, their preferences for closed and open landscapes, and their behavioral methods for fending off predators and competitors.

By about 1.8 million years ago, hominids were clearly eating a lot of grass somehow. The ratio of the stable carbon isotopes is different in leaves and fruit (C_3 plants) than it is in the African grasses (most of which are C_4 types), simply because of somewhat different photosynthesis mechanisms. Animals like gorillas that eat a lot of leaves make bone with the isotopic signature of the C_3; a grass-eating animal like a zebra or a warthog acquires the carbon-13 isotopic ratio of the C_4 types. And if someone eats a lot of meat from grazing animals, their bones too will look like the C_4 types. It may not be true at the level of behavior but at the level of atoms, you are what you eat. (And breathe—inhale deeply, and maybe you'll get an atom or so that used to reside in a *Homo erectus* brain.)

So we know that our ancestors shifted from low-grass to high-grass diets before 1.8 million years ago, and it probably wasn't because they were baking bread. By then, they were not just eating meat occasionally. They were eating a lot of it. They had probably figured out how to bring down big grass-eating animals, and with regularity. Glynn Isaac postulated that *Homo erectus* had not only attained meat-eating but transport of food and raw materials and the sharing of food. Richard Wrangham suggests that there was a major improvement in diet, perhaps involving food preparation—maybe even cooking the savory stew.

Homo erectus (also known, within Africa, as *H. ergaster*, but I'll lump them together for present purposes) was on stage and endured (in east Asia) until only 50,000 years ago.

M ANY new species of antelope appeared in Africa starting about 2.7 million years ago, adapted to increasingly arid conditions. While this gradual drying of the environment was an important player in the hominid story as well, our ancestors were particularly affected by the opportunities arising from rapid variability in climate, such as droughts and the temporary conversion of forests into grasslands. Static conditions may slowly promote efficiency, but climate dynamics can run pumps. The important point for this brief history of the mind is that the droughts can pump up the population size of any species that feeds, directly or indirectly, on grass—such as *Homo erectus.*

In a drought affecting the central population, those antelope species specialized for the drier periphery could actually expand their popula-tions at the expense of the more ordinary antelopes that needed a waterhole regularly. You can even see it in the monkeys. Ba-boons are Old World monkeys adapted to life in the more open woodlands, but they'll happily invade the forest when the com-petition allows. The same thing was likely true for hominids. Those capable of making a liv-ing on the periphery could likely live anywhere. It's another ex-  ample of a potential principle: adaptation to life on the fringes is a good setup for expanding back into the center, especially during droughts.

Grass can grow in places that lack enough rainfall for anything else, such as just south of the Sahara and in the steppes of central Asia. Fur-thermore, as I mentioned earlier, grass is the first thing to appear in the year after a fire has cleared off everything and restarted the plant suc-cession cycle of grass to bush to forest. This means that cycles of drought and fire can run a pump of sorts.

In the year following a fire, grasslands would greatly expand. Graz-ing animals can double and redouble their populations in just a few years. Their predators, having a longer time between generations, would more slowly catch up but, if the grass lasted long enough, they too would experience a temporary boom time. As brush and forests re-turn in many places, the grasslands become patchy. Thus some isolated populations of grazers and their predators would develop, only rarely encountering other populations long enough for some gene mixing.

This drought-and-fire cycle does not provide evolutionary advantages for the great apes in general, only for those such as *Homo erectus*, increasingly able to exploit herds of large grazing animals. That's one possible answer to the "Why just us?" question. (Most evolutionary arguments such as the advantages of general intelligence tend to apply equally well to other omnivores such as chimpanzees and bonobos.)

P<small>UMP</small> the Periphery is a possible principle for hominid evolution. Once there, they may find that plant succession causes grasslands to shrink in a patchy manner, perhaps stranding them in such "islands." These episodes surely emphasized the importance of such otherwise occasional virtues as cooperation, food preparation for otherwise inedible plants, and hunting efficiency.

As useful as such traits might be to a central population, the crank of Darwinian evolution turns more slowly there. Life on the fringe is, in comparison, a fast track. When deserts get enough rainfall to grow grass, it is the frontier populations that get the extra offspring surviving to adulthood.

Those who survive and thrive on the frontiers also get the chance to expand back into the central population during the next drought. Their new adaptations for grass and drought make it possible for them to make a living in more central places where the more average could not, allowing overall populations to grow. So it doesn't necessarily take frontier fighting abilities to produce that recurring theme in human history, "from periphery to center, over and over."

But, to jump ahead for a moment, the advent of herding grazing animals on a commons, rather than merely preying on them, makes one very vulnerable to theft, where the accumulation of a lifetime can be lost overnight. People become very worried about appearing weak and so violently "defend their honor" at the slightest provocation. And organized theft can also become a way of life:

> The successive waves of "barbarians" [from the steppes of northeast Asia] who overran the formerly Roman lands for more than a thousand years . . . were toughened by having lived an outdoor life in harsh, unforgiving surroundings. They were experienced in making war. They were hungry, whether for grazing lands or for plunder.
>
> —DAVID FROMKIN, 1998

So what do such large-animal predators need, compared to great apes in general? Cooperative behaviors are usually important to such predation. Indeed, even if a lone hunter kills a large antelope, it is too much meat for even a single family. The obvious strategy is to give most of it away and count on reciprocity tomorrow. Tolerated scrounging can develop into more elaborate forms of reciprocal altruism.

Sharing has a long growth curve, unlike most things that evolution operates on. You can share more things, with more people, over longer periods of time—all for additional payoffs. Human-level cooperation has come to emphasize a delayed reciprocity in which each partner risks short-term costs to achieve a long-term mutual advantage.

There are a number of ways to hunt but the one with the long growth curve is accurate throwing. Twice as far, twice as fast, twice as accurate—they are all likely to mean your family eats high-calorie non-toxic food for additional days of the month. Set pieces (like the modern dart throw or basketball free throw) are not as useful for hunting as a versatile throwing capability, able to improvise on the spot. And once you can reliably hit moving targets, things again improve.

But accurate throwing (as opposed to, say, the chimp's fling of a branch) is a difficult task for the brain. During "get set" one must improvise an appropriate-to-the-target orchestration of a hundred muscles and then execute the plan without feedback. While there are hundreds of ways to throw that would hit a particular target, they are hidden amidst millions of wrong answers, any one of which would cause dinner to run away. Planning it right the first time, rather than trying over and over, has real advantages.

Since the great apes are not noted for their planning skills, we might infer that hominid planning skills were improving out in the grasslands. It's not clear when this intensified, but it is a long road from the occasional accuracy of a chimp fling to the right-on-target high velocity skills of a baseball pitcher.

The improvement doesn't mean there was a bump developing on the skull that we might label "hand-arm planning center." Nor is there a reason to expect it to rate a "for the exclusive use of" label. In modern stroke patients, one sees a lot of overlap between hand-arm and oral-facial planning (also, and this may prove important, they both overlap with language). Though the natural selection payoff might be the hand-arm planning that orchestrates a brief ballistic movement, the same improved neural machinery is likely available for planning on longer time scales and other muscle groups. (Just remember curb cuts for wheelchairs and their free use for skateboards.)

H OMO ERECTUS promptly spread out of Africa into Asia by 1.7 million years ago. It was still in east Asia only 50,000 years ago, in the middle of the most recent ice age. *Homo erectus* was a very successful species. Why did it endure?

Perhaps they had learned to delay food consumption as well as to hunt, to prepare plant foods by pounding and soaking them first. Some think that cooking was also invented early in the *H. erectus* era. Such forms of preparation considerably expand the diet, important in hard times when the choice has become restricted. (Though Japanese monkeys can be seen to wash the sand off of food, and even to throw handfuls of grain mixed with sand into the water so as to eat what floats, the name of their game is still immediate consumption, not an intermediate product.)

Together, hunting and food preparation are probably what allowed *Homo erectus* to live in the more arid areas with only occasional trees—say, on the "shores" of the Sahara and in the steppes of central Asia.

By about 1.5 million years ago, the almost designed-looking "Acheulean" toolkit developed, the second big step up in toolmaking. Making an Acheulean handaxe required a lot more sustained effort, with an unseen goal (that flattened-teardrop shape, edged all around) held in mind. In captivity, apes can shatter rocks to get a sharp edge, and use it to cut the rope that keeps a box of bananas closed. But making something to a certain design seems to be another matter, likely requiring a great deal of tutoring and more of an attention span than apes usually have.

With its flattened-teardrop symmetry, the Acheulean handaxe has long invited cognitive explanations. It is the earliest hominid tool that seems "designed" in some modern sense. Yet for most of the "Swiss Army knife" multipurpose suite of proposed uses (defleshing, scraping, pounding roots, and flake source), an easy-to-make shape would suffice— and indeed the simpler tools continued to be made. None of these uses adequately addresses the "design aspects." Why is the handaxe mostly symmetric, why mostly flattened, why the seldom-sharp point, why sharpened all around (when that interferes with gripping the tool for pounding uses)?

Nor does a suite of uses suggest why this form could remain the same from southern Africa to western Europe to eastern Asia—and resist cultural drift for so long. The handaxe technique and its rationale were surely lost many times, just as Tasmanians lost fishing and fire-starting practices. So how did *Homo erectus* keep rediscovering the enigmatic handaxe shape, over and over for nearly 1.5 million years? Was there a constraining primary function, in addition to a Swiss Army knife collection of secondary uses?

Elsewhere I describe the handaxe's extraordinary suitability for one special-purpose case of projectile predation: attacking herds at water-holes on those occasions when they are tightly packed together and present a large, stampede-prone target. Briefly, in the beginner's version that uses a tree branch rather than a stone, the hunters hide near a waterhole. When the herd is within range, the branch is flung into their midst. The lob causes the herd to wheel about and begin to stampede. But some animal trips or becomes entangled by the branch. Because of jostling and injury by others as they flee, the animal fails to get up before hunters arrive to dispatch it.

One can imagine that tree branches were soon in short supply near waterholes. If our waterhole hominids resorted to second best, lobbing a rock into the herd's midst, it would not trip animals but it might knock one down. Because of the delaying action of the stampeding herd, this too might allow an animal to be caught. Even when you miss, the herd will be more tightly packed together on its next cautious visit to the water's edge—a sea of backs with few gaps makes it even harder to miss.

What rocks would work best? Large rocks, but also rocks whose shape had less air resistance. Most rocks tumble, but flat rocks (say, from a shale outcrop) will sometimes rotate in the style of a discus or frisbee, keeping the thin profile aligned to the direction of travel and thereby minimizing drag. Because animals will keep their distance when under heavy predation, range would become increasingly impor-tant. (Throwing farther is not the problem so much as the increased ac-curacy needed. Twice as far is about eight times as difficult.)

Hunters might also have noticed that stones with sharp edges were more effective in knocking an animal off its feet, even when not heavy. Withdrawal reflexes from painful stimuli, such as a sharp prick from an overhanging thorn tree, cause a four-legged animal to involuntarily squat. Even if the spinning stone were to hit atop the animal's back and bounce free, it might cause the animal to sit down. It is the sudden pain that is relevant, not any actual penetration of the skin. That's my theory for why handaxes are somewhat pointed (it will snag in a pushed-up roll of skin), sharpened all around, and flattened.

The handaxe would also be useful for that "Swiss Army knife" suite of secondary uses (defleshing, scraping, pounding roots, and flake source) but none of those uses tells you why it is shaped the way it is: unless damaged and reused, it is like a flattened teardrop, edged all around.

So what cognitive ability was needed by early *Homo erectus* for hand-axe design? Not much more than for shatter-and-search. Rather than being seen as an embarrassing exception to 50,000-year modernity, the handaxe can be seen—once the singular controlling use is appreciated—as having a very pragmatic shape, where deviations from the flattened teardrop are more likely to result in dinner running away. The step up to staged toolmaking (first shape a core, then knock off flakes) at 400,000 years ago is far more impressive as evidence of enhanced cognition.

So the thought processes of *Homo erectus* were surely different from what is seen in great apes—there was likely more sharing and planning—but there is still no evidence of increased creativity or art, and I'd discount the otherwise suggestive handaxe evidence as suggesting esthetic design in toolmaking.

In appearance, they were almost human. In intellect, they were likely only apes of a superior sort, not even halfway there—but with all the moves of an accomplished hunter and sophisticated gatherer.

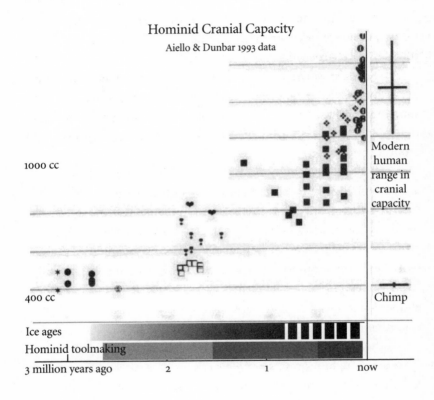

Hominid Cranial Capacity
Aiello & Dunbar 1993 data

1000 cc

400 cc

Modern human range in cranial capacity

Chimp

Ice ages

Hominid toolmaking

3 million years ago   2   1   now

5

The Second Brain Boom

What kicked in, about 750,000 years ago?

THE FIRST BRAIN BOOM started with the advent of the *Homo* spinoff, 2.5 million years back. Over the next 1.5 million years, brain size doubled. Because body size was also growing (and that alone will increase brain size), this doubling in size is not as impressive as it originally looked.

Plotting the skull sizes from all the hominid species against time, it is apparent that brain size started growing more rapidly about 750,000 years ago. And not much of this second brain boom can be attributed to a parallel enlargement in the body itself.

So what's going on here? No one knows because the data are so sparse. It's easy to produce a list of things that had to get started sometime in the last 7 million years. Indeed, there are a hundred differences between the great apes and humans and most cannot be pinpointed in time. Paleoanthropology is like a jigsaw puzzle with lots of pieces missing—and those pieces that you dig out of the ground are so "fuzzy around the edges" that they seem to fit in a number of places.

THE Australopithecines endured until about a million years ago, looking more and more heavily built, like the gorillas. When they died out, it left *Homo erectus* as the only hominid game in town. Starting at about 800,000 years ago, *Homo antecessor* was found in Spain, but no one knows whether it evolved there or in Africa—or what it was doing differently. *Homo erectus* carried on elsewhere.

While toolmaking doesn't seem to change much at this time, several other parts of the puzzle do. This is about when the ice age climate rhythms are modified. The major drivers are well known. The tilt of the earth's axis changes from 24.6° down to 22.0° and back over a cycle lasting 41,000 years. The month of the earth's closest approach to the sun, when we get about 10 percent more energy, is currently in early January but it will drift around to July in another 12,000 years or so, depending on what the other planets are doing (that's why it varies from a 19,000- to a 26,000-year cycle). And, for similar reasons, the shape of the elliptical orbit around the sun changes from rounder to more elongated over cycles that mix a minor 100,000-year and a stronger 400,000-year component. The three rhythms combine to produce a complex fluctuation.

The global amount of water tied up in ice sheets can be estimated from sea-floor cores, although this does not give a very complete picture of changing climate. Up until about 750,000 years ago, the successive meltoffs of ice were about 41,000 years apart, dominated by the tilt cycle. More recently, the period between major meltoffs has been closer to 100,000 years and the amplitude of the cycle has been greater.

Since the astronomical factors are unlikely to have changed strength or rhythm, and the astronomical 100,000-year component is relatively weak, the change is probably a matter of an alteration in how the earth resonates with the driving rhythms. For example, it takes time for ice

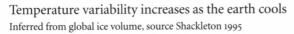

Temperature variability increases as the earth cools

Inferred from global ice volume, source Shackleton 1995

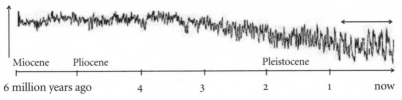

Miocene   Pliocene   Pleistocene

6 million years ago   4   3   2   1   now

sheets to become heavy enough to depress the earth's crust, and when they melt it takes time for the surface to spring back up. As for why the rhythm shifted when it did, no one knows, but the most recent reversal of the earth's magnetic poles was about 780,000 years ago (and some people think we are getting close to another reversal now).

The increase in the height of the swings suggests that other things might be going on, such as more instability of the sort associated with the big abrupt climate changes. The paleoclimate records aren't sufficiently good to see very much in detail except the last ice age, but at least one massive type of abrupt climate change, the Heinrich events associated with the collapse of the Hudson Bay ice sheet, can be seen as far back as 1.1 million years ago.

MORE demanding hunting techniques are one possibility for why brain size starts increasing faster. Perhaps techniques changed from running down prey to projectile predation. It could be a shift from side-of-the-barn throws at large targets (say, herds bunched up while visiting the waterhole) to more accurate throws involving individual prey animals at greater distances. It is only the accurate throws that make a lot of demands on the brain, compared to what great apes can do.

All throws require some planning during "get set," and so even the inaccurate throws might instill some preparatory traits that could carry over to preparing for other ballistic tasks (say, clubbing and hammering). But accurate throws really have to tune up this neural machinery.

If my Law of Large Numbers analysis in *The Cerebral Code* proves correct, you have to borrow some inexpert areas of the brain temporarily to assist the expert areas, much as the amateur audience assists the expert choir in singing *The Hallelujah Chorus*. So the "get-set" prelude involves a major amount of reassignment of cerebral resources, quite unlike most cognitive tasks. Once you start refining accuracy, however, there is that long growth curve where payoff increases with each redoubling of distance achieved with accuracy.

PROTOLANGUAGE is my other candidate for what might fit the jigsaw puzzle for this period. While the burst of creativity about 50,000 years ago was originally credited to the onset of language, the last half-century of research on language origins suggests that language is a two-step affair—or, rather, a ramp of improvements followed by a big step up to structured language. Perhaps the ramp was rising about 750,000 years ago and the step is what comes 50,000 years ago.

Protolanguage is a distinction developed by the linguist Derek Bickerton from his studies of how unstructured pidgin languages are converted by children into fully structured creole languages. You can see the same transition in the stages through which a child develops language. First come words, then word combinations that have an additional meaning besides those of the component words. Short sentences do not often require any notions of structuring to be understood, but once you try to double the number of words in a sentence, its meaning becomes quite ambiguous without some structural scaffolding that we call syntax.

Still, you can say quite a lot with two-word sentences, compared to what other animals accomplish with their one-call-one-meaning vocalizations. While the child's passion for naming things might lead us to think that nouns are the big thing, the emotional expressions of other primates are a lot closer to verbs.

Most words are a bit abstract, more categories than labels for an individual or a place. Proper nouns are considerably more difficult, it seems, than categories. For example, a brain-damaged patient might be able to name the make and model of cars in a series of photographs but be unable to pick out a picture of his own car. So the evolution of language abilities might not be "naming the creatures" so much as a series of prompts for action, equivalents of "Let's go" and "Look there."

Some such foundation is likely how protolanguage got started. Unlike the onset of syntax (more in a moment), it certainly looks as if protolanguage progress could be gradual, with no really big steps needed, just a growing vocabulary and then two words paired for a third meaning. It could have been going on for several million years, or it might have begun only 160,000 years ago with the advent of anatomically modern *Homo sapiens*. No one knows. But neocortex is all about forming new associations between concepts, and neocortical size expands more than fourfold between the great apes and modern humans. So it would not be surprising if the novel meanings for word combinations might have promoted, or profited from, a bigger brain.

Body postures communicate mood and intention (dogs communicate dozens), and arm or face posture sequences provide even more bandwidth for broadcasting your emotions and intentions. Species-specific vocalizations get a big addition from culturally defined "words" (whether signed or spoken) whose learned meaning depends much more on context.

Word combination is just another example of context dependence, but it was likely an important step. Yet with only protolanguage, you couldn't say "Who did what to whom" much faster than you could pantomime it. You'd have to make a series of short sentences rather than one compact structured sentence. Pantomime tends to be appealing as an early stage but acting it out might not have been so common early on. Pretense is involved, and playing a role (where your actions are to be interpreted as those of someone else) might not have come along until structured thought arose. Still, you can do a lot without much pretense—say, pounding on something while looking at a third party might communicate to your friend what happened in her absence, just by simple association.

It was originally supposed that coordinating hunts was a big early payoff for language—until it turned out that chimps had all the basic moves without using vocalizations. Now it is supposed that much of the everyday payoff for language has to do with socializing and sexual selection, where "verbal grooming" and gossip become important players. Again, it looks as if a gradual improvement ought to work, and that identifying starting times is likely to be less relevant than finding periods of more rapid progress.

Words are tools in some sense and extend the realm of thought beyond the here-and-now. There is one class of words that might have been particularly handy as protolanguage progressed in an era of hunting. These are the "closed class" words (so called because, unlike nouns and verbs, it is so hard to invent a new one) that serve to orient you. Some indicate relative direction (*to, from, through, left, right, up, down*) in the manner of vectors. Words such as *above, below, in, on, at, next to*, and *by* serve to orient relative to other objects. In the brain, such spatial tasks tend to involve midbrain areas such as superior colliculus and the "where" specializations of the upper parts of both parietal lobes. Because the frontal lobe tends to be involved with planning, perhaps the closed-class words for relative time (*before, after, while*, and the various indicators of tense) might live up there instead.

While there is an elementary sort of directional "structure" involved here, it is not open ended in the sense of being expandable (in the manner of nouns and verbs). The reason that it's a closed class of words is that it is about coordinates, and you only need so many words for the four dimensions of space-time, even when you add relative terms to relate several objects and their movements (*inside, beneath, alongside, atop*).

Such are not the type of recursion and nesting that constitutes the big step up to syntax—and, more generally, structured thought—with its open-ended nature. Still, the acquisition of these "little words of grammar" would have made short sentences much more versatile and hominid mental life even less like that of the great apes. With them, you could begin to order the world.

*A*ll humans do it. Gossip, schmooze, chitchat, gab, talk, tattle, rap, banter, discuss, debate, and chew the fat. Why? To exchange information, share knowledge, criticize, manipulate, encourage, teach, lie, and self-promote.

—MARC D. HAUSER, 2000

*S*eeing has, in our culture, become synonymous with understanding. We "look" at a problem. We "see" the point. We adopt a "viewpoint." We "focus" on an issue. We "see things in perspective." The world "as we see it" (rather than "as we know it" and certainly not "as we hear it" or "as we feel it") has become the measure for what is "real" and "true."

—GUNTHER KRESS AND THEO VAN LEEUWEN, 1996

*P*erhaps *surprisingly, for many animal species it is not the creative component, but rather the stabilizing ratchet component, that is the difficult feat. Thus, many nonhuman primate individuals regularly produce intelligent behavioral innovations and novelties, but then their groupmates do not engage in the kinds of social learning that would enable, over time, the cultural ratchet to do its work.*

—MICHAEL TOMASELLO, 2000

[*O*ur] *ways of knowing and core intuitions are suitable for the lifestyle of small groups of illiterate, stateless people who live off the land, survive by their wits, and depend on what they can carry. Our ancestors left this lifestyle for a settled existence only a few millennia ago, too recently for evolution to have done much, if anything, to our brains. Conspicuous by their absence are faculties suited to the stunning new understanding of the world wrought by science and technology.*

—STEVEN PINKER, 2002

6

Neanderthals
and Our Pre-*sapiens* Ancestors

Two-stage toolmaking and what it says about thought

B Y ABOUT 400,000 YEARS AGO, brain size is beginning to overlap with the modern range of brain size—the average isn't there yet, but some individuals then had brains just as large as many people of normal intelligence now have. Though *Homo erectus* carried on in most of Asia, there was a new species called *Homo heidelbergensis* in Africa and Europe, perhaps the common ancestor of both our *sapiens* lineage and that of the Neanderthals.

Some spectacular evidence for advanced projectile predation is seen about then, suggesting that less advanced forms had been around for some time before. One spear discovered in a coal mine at Schöningen, Germany, has a split shaft for hafting, suggestive of mounting sharp points. Found lying among stone tools and the butchered remains of ten horses in layers dated to about 400,000 years ago, there are also three wooden spears several meters long. They were made from the trunks of spruce trees that were about 30 years old. After removing the bark, they were sharpened at the base of the trunk, where the wood is hardest. The thickest and heaviest part of the carved shaft is about a third of the distance back from the spear tip.

So these three weren't the crude beginnings of spears, of the sort useful for thrusting and keeping troublesome scavengers at a distance. Balanced like modern javelins, they were surely thrown.

THE third big advance in crafting stones also occurred about this time, staged toolmaking. Like staged food preparation, it suggests that hominids had learned to think ahead and prepare an intermediate product.

Producing a sharp edge seems to have been the major preoccupation of toolmakers, starting 2.6 million years ago. Though random bashing works (shatter and search), it is wasteful of raw material and, if you don't happen to live in the midst of plenty, you have the motivation to make your hand-carried rock produce more sharp edges than random bashing produces. And so, early on, directed blows became common. The idea may not have been to construct a particular shape of tool, so much as to get more sharp edges per rock.

The second stage, known as the Acheulean, involved both hard hammering for the basic shapes and soft hammering for the edges. The handaxe was only the most enigmatic of the Acheulean tools. This was a gradual working of the stone to achieve a desired form, though not staging per se.

What developed was a method of staging, controlling the shape of the struck-off piece so that it had two good edges in a V-shape. The knapper first shaped a rock to have two sloping platforms like a tent. Then in the second stage, the knapper would up-end this "core" and strike in a line almost parallel to the "ridgeline." The "flakes" struck off would thus be triangular, with two sharp edges intersecting at the ridgeline. The successive flakes would get bigger as they worked down through the core. Called Levallois flakes after the Paris suburb where they were first discovered, they are also common in Africa in the same period.

W HOLE books are regularly written about Neanderthals, and no wonder. They represent "the path not taken" (that is, by the way, an advanced form of metaphor; more later) by our ancestors, yet a path that led to a species that thrived for a long time in Europe and the Middle East.

Did they lack our kind of hunting techniques? (They certainly suffered a lot more nonfatal injuries, and died at earlier ages, than comparable "paleoindian" populations, so perhaps they regularly got too close to thrashing animals, or regularly beat up on one another.) Did they lack our kind of tools? (That case was once made from the archaeological record, but now it seems they had comparable tools to Cro-Magnons where they overlapped in their habitats.) Could they make such tools, or just trade for them? Invent them, or only copy them?

Did Neanderthals come to a violent end at the hands of our ancestors? Die from imported Cro-Magnon diseases, the way smallpox cleared the way for European settlers in the Americas? Or were they simply outcompeted, slowly declining into disappearance as their hunting grounds shrank during a bad drought? (That's what evolutionary theory suggests ought eventually to happen to one of two species occupying the same evolutionary niche.) There are no answers to most such questions, but "all of the above" seems likely at different times and places.

D ID Neanderthals have our kind of language? A theme of many novels is the potential conflict between Neanderthals and modern *Homo sapiens* because one had the bleached skin of sunshine-starved Europeans and the other had the dark skins of those fresh out of sunny Africa. Cross-group romances and cross-rearing of orphans allow the novelist to show the contrasts, head to head. But the major ploy of the paleonovelist (many of whom are well informed about the anthropology) is to assume that one group had our kind of language and the other didn't.

That is indeed a key scientific question, but no one has yet found evidence that is widely persuasive on the issue of whether Neanderthals had language. Still, the effort so far shows the candidates and they say something about what had happened in the 1.2 million years between early *Homo erectus* and *Homo heidelbergensis.*

Since women and children speak at a higher pitch than adult men, thanks to their smaller size, the hearing part of our brain needs to correct for this, so that words from a small vocal tract still are recognized as the same as the words spoken from a deeper throat. Some of the vowel sounds that we make are useful as calibration signals and cause us to treat the other speech sounds in a manner appropriate to the size of the vocal tract that produced them. The upper part of the vocal tract of Neanderthals is not shaped the same as ours, judging from the curvature of the base of the skull. The argument about Neanderthal abilities has been about speech, not language per se, and has been founded on the observation that the Neanderthal throat would not have been well suited for the production of the vowels *a, i,* and *u.*

Fine, perhaps they calibrated speech sounds in some other way. Or just slowed down, the way we do when conversing with someone with a speech defect or a hearing problem. What we really want to know is how much Neanderthals and our ancestors talked and, if they could, whether they had protolanguage or, better yet, syntax.

Well, how about the size of the nerve that controls movements of the tongue? (You can infer it from the size of the hole in the base of the skull where the hypoglossal nerve exits.) It's bigger in us than in great apes or *Homo erectus.* Neanderthals are like us in this regard. But blood vessels also travel through the same holes as the nerves, and you cannot control for that, so we are not entitled to conclude that bigger holes were for bigger nerves. Worse, nerves are bundles of many nerve fibers and there are two major ways that they can get bigger: more fibers (the usual assumption, with its implication of finer control), and fibers whose diameter is greater because they have more fatty insulation. (More insulation makes them conduct faster, and also lowers the metabolic costs.)

The other line of evidence comes from the size of the nerves controlling breathing. Same results: Neanderthals are enlarged much like we are, but poor *Homo erectus* back at 1.6 million years ago had to get along with ape-sized nerves. Same caveats, too. Still, larger nerves do indeed suggest finer control of the chest muscles, and that is something that you would expect to improve with the careful modulation of breathing needed for speech—or for that matter, swimming, sustained running, and blowing at embers to keep the fire going.

None of this proves that Neanderthals indeed had our vocalization capabilities. Both lines of nerve size evidence, taken together, suggest that finer control of tongue and chest movements developed sometime between 1.6 million years ago and about 400,000 years ago when we shared a common ancestor with the Neanderthals—and that is at least consistent with a lot more vocalization.

S TAGING emerges as the central feature that says something about the mental capabilities of the hominids of this period. You have to be able to "see" a standard series of blades within the core, something like imagining sliced bread and shaping the loaf accordingly.

But, like staged food preparation, it is a routine sort of staging, passed on to an apprentice. It takes time to get good at it, just as young chimps take six years to get good at nut cracking. This sort of staging is not necessarily a new set of stages each time, the way a short-order cook can juggle an order from a whole table of people and have this unique combination all finish up at the same time.

Remember those javelins, however. That's an advanced sort of projectile predation, and surely only some of the throws were set pieces performed in a stereotyped ambush setting. So their brains were likely busy during "get set," trying to create a novel set of movement commands.

The issue is how much of that novel staging carried over to planning other things. If they could stage both toolmaking and food preparation, perhaps their life of the mind included other kinds of agendas as well, with more of an eye to the future. Maybe they added "ready" to the front end of "get set, go" just as accurate throwing and hammering prefaced the apelike "go" with a hominid "get set" phase.

To get ready, you have to set the stage with the right stuff, slowly get everything into position. Getting set requires orchestrating everything offline. And for ballistic actions, launching is entirely on automatic because feedback is too slow to modify the movement. So perhaps the evolutionary ordering is go, set, ready.

*M*orality, after all, did not enter the universe with the Big Bang and then pervade it like background radiation. It was discovered by our ancestors after billions of years of the morally indifferent process known as natural selection.

—STEVEN PINKER, 2002

*C*learly, we are not the result of a constant and careful fine-tuning process over the millennia, and much of our history has been a matter of chance and hazard. Nature never "intended" us to occupy the position of dominance in the living world that, for whatever reasons, we find ourselves in. To a remarkable extent, we are accidental tourists as we cruise through Nature in our bizarre ways. But, of course, we are nonetheless remarkable for that. And still less are we free of responsibility.

—IAN TATTERSALL, 2002

Top: Cliff overhanging modern buildings, Les Eyzies de Tayac, France
Bottom: Cro-Magnon rock shelter at base of overhang, Les Eyzies

Behaviorally modern *Homo sapiens sapiens* and their tools from about 28,000 years ago were initially found here in 1868. Similar rock shelters in the area subsequently yielded more ancient Neanderthals and flake toolmaking.

7

Homo sapiens
without the Modern Mind

The big brain but not much to show for it

I F THE APE-TO-HUMAN SAGA were a movie, we'd appear only in the final few minutes—and a silent movie might have sufficed. By the middle of the previous ice age, about 200,000 to 150,000 years ago, the DNA dating suggests that there were less heavily built people around Africa who looked a lot like us, big brains and all. And now there are some skulls from Ethiopia dated at 160,000 years ago.

If they did have language but hadn't yet made it past the short sentence stage, a silent movie might still suffice because that kind of language employs a lot of redundancy. You can often guess what's being said from the situation, the gestures, the facial expressions, and the postures. But long sentences involving syntax would surely need the soundtrack—unless, of course, structured sign language came before spoken language, always a possibility. (It's still not clear when vocalizations became such a dominant medium for the longer communications.)

But what's new with modern-looking *Homo sapiens*? That news is so minor that it gets lost in the aforementioned dramas.

THE Neanderthals came on the anthropological scene with the discovery in 1856 in the Neander valley ("tal" in the modern German spelling, "thal" in the former spelling) near Düsseldorf. The first modern human skeletons to be recognized as undoubtedly ancient were discovered in France in 1868, surely contributing to the French sense of refinement over the Germans. (English pride erupted in 1912 with the discovery of Piltdown man, but it turned out to be an elaborate fraud, uncovered only in 1949 by the invention of the fluorine absorption dating method. But at least the English had Charles Darwin to brag about.)

The "Cro-Magnons" came from a rock shelter in the Dordogne valley that was being excavated as part of the construction of a railroad. The rock shelter is nestled, like much of the town of Les Eyzies de Tayac, under the overhang of a towering limestone cliff. It can be visited today at the rear of the Hotel Cro-Magnon. The bones came from a layer, about 28,000 years old, that also yielded the bones of lions, mammoths, and reindeer. Tools were also found there in somewhat deeper layers, made in an early behaviorally modern style that archaeologists call the Aurignacian, associated with an advanced culture known for its sophisticated rock-art paintings and finely crafted tools of antler, bone, and ivory.

Many other early modern humans have been found since, mostly in Europe—but the oldest ones are from Africa and Israel at dates well before behaviorally modern tools and art appear on the scene. That's where the modern distinction comes from, of anatomically modern (*Homo sapiens*) and then behaviorally modern in addition (sometimes called *Homo sapiens sapiens*, though probably not doubly wise so much as considerably more creative).

The distinction between *Homo sapiens* and the earlier Neanderthals is based on looks, not brain size or behavior. So what makes a skull look modern? The *sapiens* forehead is often vertical, compared to Neanderthals or what came before, and the sides of the head are often vertical as well (in many earlier skulls, the right and left sides are convex). Think of our modern heads as slab-sided, like one of those pop-up tops that allows someone to stand up inside the vehicle when camping. But the back of the modern head is relatively rounded and has lost the ridge that big neck muscle insertions tend to produce in more heavily built species. Though some moderns have brow ridges up front, most do not. The face below the eyes is also relatively flat, nestling beneath the braincase, and later there is often a protruding chin. (Both likely represent a reduction in the bone needed to support the teeth, another sign that food preparation had gotten a lot better.)

The *sapiens* skeleton below the neck is less robustly built as well. Neanderthals, while often tall, had shorter forearms and lower legs, and virtually no waist because the rib cage is larger and more bell shaped. Their bones were often thicker, even in childhood when heavy use isn't an issue. Overall, the impression is that we moderns are much more lightly built than the Neanderthals, even when the same height, much as normal cars and "Neanderthal" SUVs can still fit the same parking space. Apparently we didn't have to be as tough—and it wasn't because the climate was better. If anything, the climate was worse, abruptly flipping back and forth in addition to the slower ice age rollercoaster. So something was starting to substitute for brawn.

Neanderthals were living longer than *Homo erectus* did; the proportion of individuals over 30 years old to those of young adults grew somewhat at each step from Australopithecine to *erectus* to Neanderthal; the big step in middle-age survival was when modern humans came on the scene. In an age before writing, older adults were the major source of information about what had happened before. Was this just techniques and memory for places, or was language augmenting this potential for expanding culture?

The paleoanthropologist Ian Tattersall thinks that Neanderthals had "an essentially symbol-free culture," lacking the cognitive ability to reduce the world around them to symbols expressed in words and art. Perhaps not even words, the protolanguage that I speculated about back at the 750,000-year chapter. One could make the same argument for the anatomically modern *Homo sapiens* in Africa. A lot still needs sorting out, but I'd emphasize that cognition isn't just sensory symbols; it's also movement planning, and that shows up in different ways, like those javelins in the common ancestor, 400,000 years ago.

THROWING is only one of the ballistic movements and you might wonder if hammering, clubbing, kicking, and spitting have a similar problem to solve. But they are mostly set pieces like dart throws, not needing a novel plan except for the slow positioning that precedes the action. None has the high angular velocity of the elbow and wrist that complicates timing things. With wide launch windows, it isn't such a difficult task for the brain during "get set." So I'd bet on throwing as the "get-set" task that could really drive brain evolution—particularly because it is associated with such an immediate payoff when done right.

When did spare-time uses develop for the movement planning neural machinery for throwing? Hammering was likely the earliest, though I wouldn't push cause and effect too far here. With shared machinery, you can have coevolution with synergies: better throwing might improve, in passing, the ability to hammer accurately. And vice versa.

When did the multistage planning abilities get used on a different time scale, say the hours-to-days time scale of an agenda that you keep in mind and revisit, to monitor progress and revise? That's harder but many people would note that an ability to live in the temperate midlatitudes requires getting through the months called wintertime when most plants are dormant and shelter is essential. Absent food storage, it suggests they were already able to eat meat all winter. Clearly *Homo erectus* managed all this in the Caucasus Mountains, at the same latitude and continental climate as Chicago, by 1.7 million years ago—so most of the wintertime arguments we make about hominids are applicable to much earlier dates than *Homo sapiens.*

When did secondary use spread from hand-arm movements to oral-facial ones? Or to making coherent combinations of more symbolic stuff, not just movement commands? One candidate for both is at 50,000 years ago, as that's when behaviorally modern capabilities seem to have kicked in and launched people like us out of Africa and around the world.

WHAT were the early moderns doing that was different from what came before? Is there a gradual ramp up to the behaviorally modern efflorescence? Or was there 100,000 years of "more of the same" until the transition at 50,000 years ago?

Again, that's one of the things that used to be more obvious than it now appears. Most of the archaeology came initially from sites in Europe and owed a lot (as it still does) to mining, railroad construction, and enough educated people in small towns (typically clergy and physicians, plus the visiting engineers) to take an interest in what some laborer found and report it to a relevant professor in a larger city, who might then come and investigate.

One of the early distinctions was between the Levallois "flake" technology and a finer "blade" technology. The flakes were seen at older dates and sometimes in association with bones of Neanderthals. The blades were found mostly in the more recent layers, from after Cro-Magnon humans had arrived. So blades came to be seen as part of the behaviorally modern suite of traits. People who liked to arrange things in ladders saw the older Neanderthals as producing tools that were less refined than those of the more modern types of hominid.

It sounded sensible but, thanks to a lot of hard work by the archaeologists, this hypothesis has been breaking down, as I mentioned earlier. Science is like that. Propose an explanation, and other scientists will take it as a challenge. Some scientists will even undertake to disprove their own hypothesis. (You want to find the flaws yourself, and go on to find a better explanation, before others do it for you.)

A more careful look at the European sites showed that they don't reliably segregate into a Neanderthals-and-flakes cluster versus a moderns-with-blades cluster. Worse, the blade technology has been found in Africa at about 280,000 years ago, which is back even before anatomically modern humans were on the scene. So behaviorally modern Africans might have brought the blade technology along with them when they emigrated to Europe, but the Neanderthals seem to have known a good thing when they saw it.

Also, blades can be seen as just a refinement of the staged toolmaking needed for flakes, not another big step up in toolmaking technology. Instead of preparing a core to have the tentlike sloping sides meeting at a ridgeline, what produces the triangular shape of the flake, the core for a blade tends to have one end flattened, so that there is a nice right-angle cliff. Like jumping on the edge of a cliff (what boys sometimes try on a dare, jumping to safety as the leading edge collapses), the toolmaker uses a pusher piece (say, a small bone) to stand atop the core's "cliff edge." A tap on this punch with a rock hammer will then shear off a blade.

Using a punch ensured that the pressure focused on a narrow edge. This technique allowed a thin blade to be sheared off—but it is not uniformly thick. Instead it often thins down to one sharp edge. It can be grasped in the manner that we would hold a single-edged razor blade. Or it can be mounted in a handle, perhaps held in a split stick. Because a whole series of thin blades could be sheared off from the prepared core, it was quite efficient in conserving raw material.

Other things were changing in this period, but it is hard to tell because *Homo sapiens* population densities were so low in Africa. They were spread even thinner than historic hunter-gatherers. In southern Africa, they were eating eland, a large antelope adapted to dry country,

Obsidian core and blades. National Museum of Kenya, Nairobi

quite regularly—but they were not often eating the more dangerous prey like pigs and cape buffalo, which can be quite aggressive toward hunters. They generally didn't fish. They didn't transport stone over long distances, even though making blades. When they buried their dead, there were no grave goods or evidences of ceremony. Campsites did begin to show some evidence of organization.

So this isn't just toolmaking. Anatomically modern *Homo sapiens* of Africa were not conspicuously successful like the behaviorally modern Africans that followed. They certainly didn't leave much evidence for a life of the mind. Perhaps archaeologists will dig in the right place and change this view, but a half-century of testing the hypothesis— that the bloom in creativity came long after *Homo sapiens* was on the scene—has led to the view that mere anatomical modernity was not the big step.

The archaeologist Richard Klein summarizes the data this way:

[The] people who lived in Africa between 130,000 and 50,000 years ago may have been modern or near-modern in form, but they were behaviorally similar to the [European] Neanderthals. Like the Neanderthals, they commonly struck stone flakes or flake-blades (elongated flakes) from cores they had carefully prepared in advance; they often collected naturally occurring pigments, perhaps because they were attracted by the colors; they apparently built fires at will; they buried their dead, at least on occasion; and they routinely acquired large mammals as food. In these respects and perhaps others, they may have been advanced over their predecessors. Yet, in common with both earlier people and their Neanderthal contemporaries, they manufactured a relatively small range of recognizable stone tool types; their artifact assemblages varied remarkably little through time and space (despite notable environmental variation); they obtained stone raw material mostly from local sources (suggesting relatively small home ranges or very simple social networks); they rarely if ever utilized bone, ivory, or shell to produce formal artifacts; they buried their dead without grave goods or any other compelling evidence for ritual or ceremony; they left little or no evidence for structures or for any other formal modification of their campsites; they were relatively ineffective hunter-gatherers who lacked, for example, the ability to fish; their populations were apparently very sparse, even by historic hunter-gatherer standards; and they left no compelling evidence for art or decoration.

So it's not just Neanderthals—people who looked like us were not exhibiting much in the way of what we associate with a versatile intelligence.

CLEARLY, having a big brain is not sufficient to produce spectacular results. It must have taken something more.

The Neanderthal brain is actually somewhat bigger on average than the early modern brain (though, when you correct for the larger Neanderthal body, the difference seems minor). But that's just the final example of why bigger brains aren't necessarily smarter. Brain size earlier increased over many periods when there wasn't much progress in toolmaking. Whatever was driving brain size must have been some-

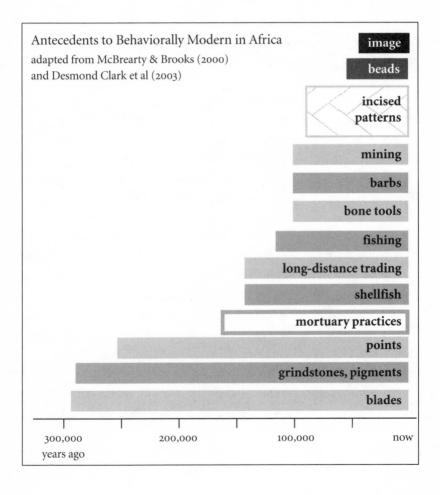

thing that didn't leave durable evidence much of the time—say, social intelligence—perhaps even something that didn't generally improve intelligence.

For example, when you approach grazing animals, you can get a lot closer if the animals haven't been spooked recently by a lot of previous hunters. There is a version of this in evolutionary time too, as the hunted animals that spook easily are the ones that reproduce better. Hundreds of generations later, the hunters have to be more skillful just to maintain the same yield. (When hominids made it out into Eurasia, the grazing animals there were surely naïve in comparison to African species, which had coevolved with hunting hominids.) Lengthening approach distances, as African herds evolved to become more wary since early *erectus* days, would have gradually required more accurate throws (just double your target distance and it will become about eight times more difficult). This is the Red Queen Principle, as when Alice was told that you had to run faster just to stay in the same place. The Red Queen may be the patron saint of the *Homo* lineage. Even if they had no ambition to get better and better, the increasing approach distances forced them to improve their technique.

That's just one example of things that, like protolanguage, are invisible to archaeology but that could have driven brain reorganization (and with it, perhaps brain size). So while the early moderns had a lot going for them, I suspect that it didn't yet include a very versatile mental life. They may have had a lot of words, but not an ability to think long, complicated thoughts. Still, they might have been able to think simple thoughts at one remove.

U NDERSTANDING another as an animate being, able to make things happen, is not the same thing as ascribing thoughts and knowledge to that individual. Babies can distinguish animate beings from inanimate objects quite early, but it is much later, usually after several years' experience with language, that they begin to get the idea that others have a different knowledge base than they have.

A simple way of testing this with a child is to use a doll named "Sally." Both the child and Sally observe a banana being placed inside one of two covered boxes. Then Sally "leaves the room" and the experimenter, with the child watching, moves the banana to the other box and replaces the lids. Now Sally "returns to the room." The child is asked where Sally is going to look for the banana. A three-year-old child is likely to say that Sally will look in the box where the banana currently is. A five-year-old will recognize that Sally doesn't know about the banana being moved in her absence and that she will therefore look in the original box. Being able to distinguish between your own knowledge and that of another takes time to develop in childhood.

Still, even chimpanzees have the rudiments of a "theory of mind," which is what we call this advanced ability to understand others as having intentions, goals, strategies, attention that can be directed, and their own knowledge base. Being able to put yourself in someone else's shoes, one setup for empathy, is not an all-or-nothing skill. Though this "theory of [another's] mind" has an aspect of operating at one remove—and thus of having some semblance of structured thought—it may have evolved before the structured suite that I will discuss in the next chapter at some length.

M IRROR neurons are in style as a candidate for the transition but, while it is exciting stuff, I suspect that "see one, do one" imitation wasn't the main thing.

Mirroring is the tendency of two people in conversation to mimic one another's postures and gestures after some delay. Cough, touch your hair, or cross your legs and your friend may do the same in the next minute or so. Many such movements are socially contagious but they are usually standard gestures, lacking in novelty.

While mirroring reminds most people of imitation, imitation actually suggests something more novel than triggering an item of your standard repertoire of gestures. But it does represent a widespread notion that observation can be readily translated into movement commands for action ("see one, do one") and that's what lies at the heart of the neurobiologists' excitement over this.

It's a neat idea, though there are some beginners' mistakes to avoid. For one thing, "monkey see, monkey do" is not as common in monkeys as the adage suggests. Not even in chimpanzees does it work well. Michael Tomasello and his colleagues removed a young chimpanzee from a play group and taught her two different arbitrary signals by means of which she obtained desired food from a human. "When she was then returned to the group and used these same gestures to obtain food from a human, there was not one instance of another individual reproducing either of the new gestures—even though all of the other individuals were observing the gesturer and highly motivated for the food."

Chimps do not engage in much teaching, either. No one points, shows them things, or engages in similar nonlanguage ways that humans interact with children. There isn't much instructive joint attention, outside of modern humans.

IF the brain is going to imitate the actions of another, it will need somehow to convert a visual analysis into a movement sequence of its own. Perhaps it will have some neurons that become active both when performing an action—say, scooping a raisin out of a hole—and when merely observing someone else do it, but not moving. Such neurons have indeed been found in the monkey's brain, and in an area of frontal lobe that, in humans, is also used for language tasks.

Yet it is easy to jump to conclusions here if you aren't actually engaged in the research. These "mirror neurons" need not be involved in imitation (the sequence of "see one, do one") nor even involved in planning the action. Every time someone mentions mirror neurons as part of a neural circuitry for "see one, do one" mimicry, the researchers will patiently point out that what they observe could equally be neurons involved in simply understanding what's going on.

Just as in language, where we know that there is a big difference in difficulty between passively understanding a sentence and being able to actively construct and pronounce such a sentence, so the observed mirror neurons might just be in the understanding business rather than the movement mimicry business. But the videos of mirror neurons firing away when the monkey is simply watching another monkey, and in the same way as when the monkey himself makes the movements, have fired the imaginations of brain researchers looking for a way to relate sensory stuff with movement planning and inspired experiments designed to tease apart the possibilities.

It's an exciting prospect for language because of the motor theory of speech understanding, put forward by Alvin Liberman back in 1967. Elaborated by others in the last few decades, it claims that tuning in to a novel speech sound is, in part, a movement task. That's because categorizing certain sounds ("hearing" them) may involve subvocal attempts to create the same sound. The recognition categories seem related to the class of movement that it would take to mimic it. More and better imitation, perhaps enabled by more "mirror neurons," might be involved in the cultural spread of communicative gestures and perhaps even vocalizations.

Humans are extraordinary mimics, compared to monkeys and apes, and it surely helps in the cultural spread of language and toolmaking. So did we get a lot more mirror neurons at the transition? Or did an already augmented human mirroring system merely help spread another biological or cultural invention in a profound way, much as it might have spread around toolmaking, food preparation, and protolanguage?

Mirroring had likely been aiding in the gradual development of novel vocalizations and words for some time before the transition to behaviorally modern. Anything that helped humans more rapidly to acquire new words could have sped up the onset of widespread language. But you don't want to confuse a mechanism that amplifies the effect of the real thing with, well, the real thing itself.

F RAMING is only protostructure and it is a commonplace of perception, shared by the other animals and surely by *Homo sapiens* back before the transition. The surroundings make a big difference. The spots are the same shade of gray, but the square black surround lightens our perception of the surrounded spot. In the moon illusion, the moon looks a lot larger when framed near the horizon.

You can easily make a symmetrical face look crooked. You can distort size and shape, as is nicely shown in Roger Shepard's "Turning the Tables" illusion. The two tabletops are actually the same size and shape but the legs and drop-shadows make all the difference in how you judge them. (Go ahead, measure them. If that doesn't convince you, make a cutout of a tabletop and reposition it over the other one.)

Similar effects happen in cognition, where we call it "context." You interpret the word *lead* quite differently depending on whether the context is a pencil, a leash, a heavy metal, a horse race, or a journalist's first paragraph.

In movement planning—the other half of cognition but often ignored by people who write on the subject—the reports of the present position of your body serve to frame the plan, as does the space around you (say, your target if hammering or clubbing). The tentative movement plan itself frames the next round of perception. We frame things in sequence, as in our anthropological preoccupation with what comes first and what follows.

And when we get to language, some of those closed-class words such as *left, below, behind, before* serve to query for elements of the frame in space and time. Framing continues to be a major player at higher cognitive levels, as we will see in the creativity chapter.

T HE big step up might involve the evolutionary equivalent of tacking on another of Piaget's developmental stages seen in children. Still another formulation has to do with augmenting "primary process"

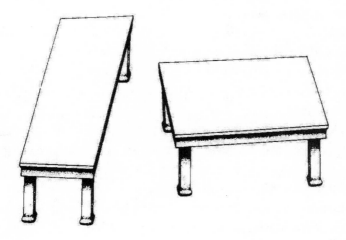

in our mental lives. The notion is that it got an addition called "secondary process." The distinction was one of Freud's enduring insights, though much modified in a century's time.

Primary process connotations include simple perception and a sense of timelessness. You also easily conflate ideas, and may not be able to keep track of the circumstances where you learned something. Primary process stuff is illogical and not particularly symbolic. You consequently engage in displacement behaviors, as in kicking the dog when frustrated by something else. There is a lot of automatic, immediate evaluation of people, objects, and events in one's life, occurring without intention or awareness, having strong effects on one's decisions, and driven by the immediacy of here-and-now needs. While one occasionally sees an adult who largely functions in primary process, it's more characteristic of the young child who hasn't yet learned that others have a different point of view and different knowledge.

Secondary process is now seen as layered atop primary process, and is much more symbolic. Secondary stuff involves symbolic representations of an experience, not merely the experience itself. It is capable of being logical, even occasionally achieves it. You can have extensive shared reference, the intersubjectivity that makes institutions (and political parties) possible.

Modern neuropsychological tests of frontal lobe function usually look at an individual's ability to change behavioral "set" in midstream. For example, the person is asked to sort a deck of cards into two piles, with the experimenter saying yes or no after each card is laid down. Soon the person gets the idea that black cards go in the left pile, red suits in the right pile. But halfway through the deck, the experimenter changes strategy and says yes only after face cards are placed in the left pile, numbered ones in the right pile. Normal people catch on to the new strategy within a few cards and thereafter get a monotonous string of affirmations. But people with impaired frontal lobe function may fail to switch strategies, sticking to the formerly correct strategy for a long time. This flexibility may be part of the modern suite of behaviors, or it might come a bit earlier.

PEOPLE who have only protolanguage, not syntax, provide some insight about the premodern condition. What might it have been like, to be anatomically modern, with a sizeable vocabulary—but with no syntax, no ability to quickly generate structured thoughts and winnow the nonsense from them?

The feral children and the locked-in-a-closet abused children are possible models, but they generally have a lot wrong with them by the time they are studied for signs of higher intellectual functions. A better example is the deaf child of hearing parents where everything else about the child's rearing and health is usually normal except for the lack of exposure to a structured language during the critical preschool years. (Deaf children exposed to a structured sign language in their early years via deaf parents or in deaf childcare will pick up its syntax just as readily as hearing children infer syntax from speech patterns.) The average age at which congenital deafness is diagnosed in the United States is when the child is three years old—which means that some aren't tested until they start school.

Certainly, one of life's major tragedies occurs when a deaf child is not recognized as being deaf until well after the major windows of opportunity for softwiring the brain in early childhood have closed. Oliver Sacks' description of an eleven-year-old deaf boy, reared without sign language for his first ten years because they mistakenly thought he was mentally retarded, shows what mental life is like when you can't structure things:

> Joseph saw, distinguished, categorized, used; he had no problems with perceptual categorization or generalization, but he could not, it seemed, go much beyond this, hold abstract ideas in mind, reflect, play, plan. He seemed completely literal—unable

to juggle images or hypotheses or possibilities, unable to enter an imaginative or figurative realm. . . . He seemed, like an animal, or an infant, to be stuck in the present, to be confined to literal and immediate perception.

That's what lack of opportunity to tune up to the structural aspects of language can do. So a child born both deaf and blind has little opportunity to softwire a brain capable of structured consciousness. (But what about Helen Keller and her rich inner life? Fortunately she wasn't born blind and deaf, as the usual story has it. She probably had 19 months of normal exposure to language before being stricken by meningitis. By 18 months, some children are starting to express structured sentences, showing that they had been understanding them even earlier. Helen Keller probably softwired her brain for structured stuff like syntax before losing her sight and sound.)

So the premodern mind might well have been something like Joseph's, able to handle a few words at a time but not future or past tense, not long nested sentences. A premodern likely had thought, in Freud's sense of trial action. But without structuring, you cannot create sentences of any length or complexity—and you likely cannot think such thoughts, either.

Imagine people like Joseph but speaking short sentences instead of signing them—yet having no higher cognitive functions, "unable to juggle images or hypotheses or possibilities, unable to enter an imaginative or figurative realm." That's what *Homo sapiens* might have been like before they somehow invented behavioral modernity.

We get tremory
In this world with no memory
Life makes only partial sense
Knowing only the present tense.

—JOANNE SYDNEY LESSNER AND JOSHUA ROSENBLUM,
FROM THE LIBRETTO OF "EINSTEIN'S DREAMS"

Alfred Russel Wallace figured out natural selection in 1858 on his own (except for several close friends, Darwin had kept quiet about it after he figured it out in 1838). But Wallace later had a great deal of trouble with whether humans could be explained by the evolutionary process as it was then understood.

He couldn't understand how physiology and anatomy were ever going to explain consciousness. (Some progress has finally been made in recent years.)

And he said that a "savage" (and Wallace had a high opinion of them) possessed a brain apparently far too highly developed for his immediate needs, but essential for advanced civilization and moral advancement. Since evolution proceeds on immediate utility and the savage's way of life was the way of our ancestors, how could that be? It sounds like preparation, evolution "thinking ahead" in a way that known evolutionary mechanisms cannot do. (Still a troublesome question; the next chapter will tackle it.)

They are both excellent observations, ones that have troubled many scientists and philosophers since Wallace stated them so well. Wallace went on to conclude, however, that some "overruling intelligence" had guided evolution in the case of humans. "I differ grievously from you," Darwin wrote to Wallace in April 1869, "and I am very sorry for it." Darwin, like most neuroscientists and evolutionary scientists today, would rather leave the question open than answer it in Wallace's manner. "I can see no necessity for calling in an additional . . . cause in regard to Man," Darwin wrote to Wallace.

A mystery, Daniel Dennett reminds us in his book *Consciousness Explained*, "is a phenomenon that people don't know how to think about—yet." To the people who actually work on Wallace's two problems, they are no longer the mystery they once were. We have great hopes of eventually proving Ambrose Bierce wrong:

Mind, n. A mysterious form of matter secreted by the brain. Its chief activity consists in the endeavor to ascertain its own nature, the futility of the attempt being due to the fact that it has nothing but itself to know itself with.

—*THE DEVIL'S DICTIONARY*

I say this, recognizing that no one person's mind can do the job of understanding mind and brain. But collectively and over time, science manages amazing feats.

We may not know all the answers about how the modern mind emerged from the ancestral mentality, but it looks as if it can be understood without calling in mysterious forces.

We human beings, unlike all other species on the planet, are knowers. We are the only ones who have figured out what we are, and where we are, in this great universe. And we're even beginning to figure out how we got here.

—DANIEL C. DENNETT, 2003

There is no step more uplifting, more momentous in the history of mind design, than the invention of language. When Homo sapiens *became the beneficiary of this invention, the species stepped into a slingshot that has launched it far beyond all other earthly species in the power to look ahead and reflect.*

—DANIEL C. DENNETT, 1996

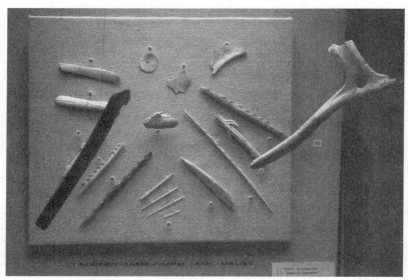

Top: A copy of the polychrome cave painting at Grotte de Font de Gaume, France
Bottom: Upper Paleolithic bone tools, Les Eyzies de Tayac, France

8

Structured Thought Finally Appears

*The curb-cut principle
and emerging higher intellectual function*

I

T'S ONLY SOMETIME IN the last 50,000 years that the archaeologists see the type of creativity that we associate with cave paintings, sewing needles, decorative carvings, pendants, and beads. Most appear on the scene only during the last half of the last ice age—which would be the last minute of a two-hour-long "up from the apes" movie.

The cave paintings speak, as Richard Leakey says, "of a mental world we readily recognize as our own." You can't make such a statement for anytime earlier. The Chauvet cave in France has fully representational paintings (even with perspective) dated to about 35,000 years ago, even earlier than that Cro-Magnon rock shelter.

This makes it look like anatomically modern *Homo sapiens* about 100,000 years ago wasn't even halfway to us. It's more like there was a jump from one-quarter to three-quarters. So what's behind this big transition? "Symbolic stuff kicked in 50,000 years ago, and that led to all manner of thoroughly modern behaviors." That's the usual explanation for the Modern Transition, and it's not a bad one.

You may have noticed, however, that I have avoided talking about symbols. That's because I consider symbols an inadequate summary, not because I think the explanation "wrong." Here's the OED on *symbol*:

> 2. Something that stands for, represents, or denotes something else (not by exact resemblance, but by vague suggestion, or by some accidental or conventional relation); esp. a material object representing or taken to represent something immaterial or abstract, as a being, idea, quality, or condition; a representative or typical figure, sign, or token.

But at root, it's an association between dissimilar things, much like Pavlov's dog salivating for the bell in addition to dinner itself. The important aspect for language purposes appears when the symbol stands in for something compounded of many other things, especially at a higher level of organization (more later) than mere objects and simple actions—say, the notion of a Rain God or an ecosystem.

If one word has to suffice to summarize a broad collection of new abilities—perhaps a bad idea to start with—I'd certainly pick a different word than the one popular throughout the twentieth century. With

notable exceptions (do read Terry Deacon's *The Symbolic Species*), when people latch on to the idea of symbolic stuff, they stop thinking about the parts and pieces—and how they are coherently assembled at different levels of organization. People go away thinking that they've achieved something by labeling all of it "symbolic" and may just argue in a circle.

Looking at the same material through the eyes of process, how neural circuits turn one thing into another, is more satisfying—and it enables you to spot more candidates for the big step up.

That there is a stand-in is not what needs emphasis. Human minds are indeed more capable of both broad categories and fine distinctions but what's important here is that the referent is a complicated compound, often framed or structured. Yet what I am calling "structured stuff" goes well beyond framing. Structuring really makes long sentences fly, and likely complicated thoughts as well.

WHAT else comes before structured language? So far I've discussed some things that might be called protostructure: staging intermediate products in food preparation and toolmaking, framing, and theory of mind. I've mentioned imitation as an amplifier. I mentioned protolanguage earlier, but let me now spell out what Derek Bickerton means by it.

This unstructured language is what you see in toddlers, speakers of pidgins (the shared vocabulary between people who lack a common language, spoken without grammar), and in some stroke patients with aphasia. It is often heavily augmented with gestures. Most of all, the sentences are short in the manner of what two-year-olds produce— you can guess the meaning without any help from word order or inflections. This is the level of production that can be achieved in tutored bonobos like Kanzi and Panbinisha (they can do somewhat better at understanding than with production, much as I do with my rusty agrammatical German). A lot can be inferred from context.

Long sentences, however, are simply too ambiguous without some mutually understood conventions about internal structuring into phrases and clauses. A clause (it contains a verb) may be embedded in a phrase and vice versa. The structuring conventions that help you figure out "who did what to whom" are called syntax and each dialect has a different way of doing things. There are five other permutations of the subject-verb-object word order used in English declarative sentences, and all can be found in some language around the world. Many languages convey structure with the aid of word endings that mark the role that the word is meant to play in that particular sentence; for example, *–ly* is an English ending that usually means that the word is going to modify a verb.

"Universal grammar" is simply the tendency of all human groups to draw from a restricted set of structuring possibilities; not every structuring scheme appears to be possible, and that restriction likely says something about the limitations of the human brain. For example, you can move one word out of place—say, in making "*What* did she give to him?" out of the standard order "She gave *what* to him?"—but two things cannot be moved in the same sentence without making it hopelessly ambiguous.

> The human faculty of language appears to be organized like the genetic code—hierarchical, generative, recursive, and virtually limitless with respect to its scope of expression. . . . Most current commentators agree that, although bees dance, birds sing, and chimpanzees grunt, these systems of communication differ qualitatively from human language. In particular, animal communication systems lack the rich expressive and open-ended power of human language (based on humans' capacity for recursion).
>
> —MARC HAUSER, NOAM CHOMSKY, TECUMSEH FITCH, 2002

SYNTAX is the best-studied case of structured thought, one of the candidates for what "paid for" the rest, and certainly the earliest one to appear in modern childhood (kids pick one up between 18 and 36 months). Once you have a syntax, you can convey complicated thoughts. And the acquisition of syntax likely tunes up the brain to do other structured tasks.

The step up to syntax is indeed a big one, as it involves such things as recursion. In saying "I think I saw him leave to go home," you are nesting four sentences like the Russian babushka dolls. Structuring helps you discover the meaning of the sentence (the mental model in the speaker's mind) but it isn't the meaning itself. Chomsky's famous sentence to show the independence of syntax from semantics, "Colorless green ideas sleep furiously," may not make any sense, but it doesn't ring any structural alarm bells in our minds in the manner of "Colorless ideas furiously sleep green."

Syntax is not necessarily a tree, though people who like maps are quite attracted to the technique of diagramming a sentence using branch points with only two daughter branches. At the heart of the structured sentence are clauses (easy to spot because each has a verb) and phrases. It turns out that they can nest inside one another. "John gave a present to Mary in the park last night" has only one verb but you can augment it by substituting "the present his sister bought" clause for "a present."

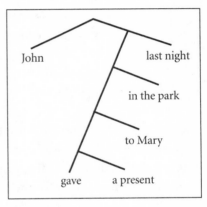

If you're not a map person, you might like argument structure better. This complementary way of looking at structure emphasizes the sentence as a little drama in which words play roles. A noun might be the actor or the recipient—you have to figure it out from clues. Some words, especially verbs, prompt you to find other words that go with them. But not just any word. It has to be a word that can play the particular role.

In any of its equivalents in the world's languages, "give" is a verb with three roles to fill (there can also be associated nouns filling optional roles such as time and place, but give needs three nouns to fill its mandatory roles). You need to find a noun that can play the role of the giver ("John"). Another noun for the thing given ("a present"). And a third noun that can play the recipient. Because an inanimate object ("present") cannot be an actor, you can immediately eliminate one role possibility.

Try to speak a sentence without one of the three essential nouns and your listener will be perplexed and go in search of the missing word. A missing subject probably means "you" as in the imperative form, "Give me that." But a billboard with a fragmentary sentence "Give him . . . " will cause a double take (exactly what the ad agency was trying to achieve) while you search for the pictorial equivalent of the object to be given.

The so-called intransitive verbs such as "sleep" may not require much in the way of a supporting cast ("He slept."), but you can always add optional phrases ("on the sofa" or "during the lecture") or adverbs such as "quietly" or "poorly." Some languages such as Latin rely heavily on such "inflections"; word order may not matter very much. In English, many inflections have disappeared in recent centuries; word order is often your main clue about what roles to assign, as in the subject-verb-object order of a simple declarative sentence. But word order is only one clue to structure, not the main thing.

The basic idea of argument structure, and of the lexicon in Chomsky's more recent minimalist grammar, is that each word has some possible roles that form part of its mental baggage. You simply know them from experience. You can't "break the blanket" because a blanket doesn't have the right attributes to be breakable. You can tear it or burn it or ruin it, but not break it.

Presented with a sentence to analyze, you try to fit the pieces together in a way that makes sense. Every required role must be filled, and no words should be left over once you have guessed the optional roles—your mental model of the sentence now hangs together. With the meaning thus extracted, you move on to the next sentence and its coherence-finding game. If a sentence is incoherent, we ask for a repeat or just accept a reasonable candidate and move on. If the sentence seems to have two perfectly reasonable alternative interpretations that leave us undecided, it may be humorous enough to make us laugh.

The verb is usually the starting point of an analysis as it tells you how many obligatory roles must be satisfied by the other words of the sentence. Some like "sleep" have only one role to fill, others such as "bring" and "give" have three; "bet" takes four as you also have to specify the condition of the bet. So understanding a sentence with syntax is like solving a jigsaw puzzle.

Language is all about taking a mental model of relationships in your mind—say, "who did what to whom"—and, via gestures or speech, getting someone else to guess exactly what you are thinking. Language can get across a model of relationships even to someone who doesn't share the context, even models of relationships that haven't happened yet, even about abstract concepts like washing machines that eat socks.

Language only needs to be good enough. We often don't need to finish sentences, because the rest can be guessed so easily. We just need enough hints to allow us to guess correctly most of the time. The task of language is to convey such relationships in an open-ended way, being able to convey novel amalgamations and have the listener effortlessly get the right idea. If you are sticking to simple relationships, unstructured protolanguage may suffice. But if you want to speak a long sentence, the ambiguity is a killer unless you can spot all of those prompts for frames and roles that we call grammar and syntax.

Syntax is our best-studied case of structured thought, given how many ways there are to do the structuring job and how many schemes seem impossible, not used in any human language. But syntax is not the only type of structured thought, and perhaps not even the first one to evolve.

Other structured aspects of thought are multistage planning, games with rules that constrain possible moves, chains of logic, structured music—and a fascination with discovering hidden order, with imagining how things hang together. This structured suite likely enabled the giant step up to the modern mind of *Homo sapiens sapiens*. Let me unpack what I mean by all the structured stuff.

Planning is not what a squirrel does as winter approaches. Earlier I made that distinction in the context of physiological mechanisms. Now note that from the evolutionary perspective, there is nothing novel about winter coming, even for a young squirrel. It has happened every year to every squirrel for as long as there have been squirrels in the temperate zone, and said squirrel comes from an unbroken line of squirrels that survived winter's dormancy of food resources. Like mating, nest building, and nurturing behaviors, food hoarding is too important to be left to learning or innovation. Planning pertains to novel situations, not learning or instinct.

Note that instincts are another nice example of framing: when a naïve animal is placed in an important setting for the first time, out pops an intricate never-experienced-before behavior like mating or nurturing. "Context, context, context" is what those real-estate agents really mean by their mantra about the three most important considerations, "location, location, location."

Behaviorally modern humans plan for things that have never happened before, and in terms of choices. ("Well, obviously if we go to the country this weekend, we can't go to the baseball game Saturday night.") Foresight often involves contingencies, another type of structure. ("We can go to the country this weekend unless I have to work Saturday, in which case maybe we can go to a movie on Sunday.") And by the time children reach school age, we start holding them responsible for having some foresight, and in a way that we do not apply to younger children or pets. (In the immortal words of my mother, "Well, you should have thought about that before you did it!")

It is useful to imagine a version of structured thought that is too slow for repartee. It might be handy when you have time to think about things overnight, and so it can influence agendas and contingent planning—but even if you could speak a novel sentence aloud, no one might be able to interpret it without thinking about it overnight. We usually assume spoken syntax comes first in evolution as it does in childhood, but overnight contemplation for planning could have been an early payoff for structured thought, even before language.

Chains of logic, like those multistage novel plans, are considerably more difficult to handle than the simple forms of logic seen in other animals.

Logical trains of inference allow us to connect remote causes through intermediate stages to present effects. The basic element may be the primitive two-stage "after this therefore because of this" attribution, but reasoning in long chains is something at which we excel. A supposed chain is, in reality, often a web instead, but the notion of being impelled down a path is very strong in us and it's much more difficult for us to think about multiple causation. Most of the things that happen in the world have multiple causes, of course, so we make a lot of errors of attribution.

When not all the elements are clear, we have a propensity to guess at chains of causation. This is very useful in doing science. You can, of course, fool yourself very easily, which is why it is so important to keep track of what is provisional or pretense and what is well established.

Games have made-up rules that you have to consult before making your move. Hopscotch and dance may have elements of play but they are also flexibly structured in ways that constrain choices. But what I have in mind here is something with an arbitrary framework of allowed moves, against which possible moves must be checked before acting.

Indeed, once four-year-olds have the ability to say "who did what to whom," they love to keep track of the actions of others (and report deviations to the person in charge). It becomes a game.

Narrative is closely related to framework-checking because we develop some standards for a good story—and not just epics but the everyday multifaceted stories, such as what we did for lunch. There's almost a "script" (with whom, where, what eaten, what discussed, and so forth). If some part is missing, we often inquire. So narrative is pattern on an even longer time scale than a sentence and it often has some blanks to fill in, just as in those "give him" sentences.

I suspect that logical chains grow out of small-scale storytelling. Someone pouring coffee into a cup provides a small story with familiar parts (called *image schemas*), the coffee *pouring* from one container, *flowing* along a *path* and then being *contained* by a second object.

As Mark Turner points out in *The Literary Mind*, partitioning the world into object categories also involves partitioning the world into small stories: catching a ball, throwing it, sitting on a chair, drinking the coffee. Many animals can do some of this, but modern humans can weave small stories into considerable narratives. We use such chains to evaluate the wisdom of possible actions, to plan better ones, and when it all hangs together well enough to connect the underpinnings, we say that we have "understood" or "explained" things.

Music can be simply patterned as in rhythm and melody, and it can additionally be structured as in harmony. Multivoiced music is what you get into with singing a fifth or an octave above someone else. It is particularly impressive when one person can manage both, as when the left hand plays a different melody from the right hand.

In Western music, this is not much more than a thousand years old. The counter melodies of the baroque, which Bach elaborated from church music that had already proved its emotional appeal in plainchant, are only a few centuries old. The philosopher Karl Popper said that the development of multivoiced music was "possibly the most unprecedented, original, indeed miraculous achievement of our Western civilization, not excluding science."

Rhythm itself may be much older and the solo voice might be an outgrowth of ancient storytelling techniques, the melody (and alliteration and rhyme) used as an aid to memory about what comes next. Music may elaborate social cohesion (marching music as a technique of the warmonger). But there may also be individual advantages to be gained from showing off unusual abilities, not only musical abilities but virtuoso performances of other sorts (intricate dances, bower building, blindman's buff). When females choose males (rather than males excluding one another from access to females), genetic fitness tends to be judged by just such complex behavioral proxies.

Discovering hidden patterns is seen in music, jigsaw puzzles, and doing science. We take great pleasure in "getting it." We love to see patterns emerging from seeming chaos, whether in doing a crossword puzzle or in doing science. Coherence-finding is probably part of the source of our musical pleasure in listening to the left hand's rhythm interacting with the right hand's melody.

As with those fill-in-the-blank test sentences, we're always guessing about missing parts, trying to make wholes out of fragments. Beyond-the-apes intelligence seems to be about making a guess that discovers some new underlying order—finding the solution to a problem or the logic in an argument, happening upon an appropriate analogy, creating a pleasing harmony or a witty reply, correctly predicting what's likely to happen next.

Indeed, you routinely guess what comes next, even subconsciously. That's why a joke's punch line or a P.D.Q. Bach musical parody brings you up short—you were subconsciously predicting and were surprised by the unanticipated ending.

Note that music beyond rhythm, planning beyond the predictable seasons, and the logical chaining of ideas are all things that involve the novel, not just learned repetitions like singing the national anthem. Novel structured stuff, with its search for coherence, is what we usually call "higher intellectual function."

The structuring that I have in mind for higher intellectual functions is not just a simple chain of events or intermediate products. It is more like a symphony—and that reminds me of another important symphony that the brain had been producing for a million years before intellect.

I call them a "structured suite" because I suspect that they share a lot of the same neural machinery in the brain, one of the reasons why some functions might come (and go, in strokes or senility) as a package deal. Perhaps the mental machinery for structuring is shared in part with some nonintellectual functions as well.

Accurate throwing (not just flinging, which many chimps do, but practicing to hit smaller and smaller targets) is not usually a set piece like a dart throw or basketball free throw where the idea is to perform the action exactly the same way as your hard-earned standard. Throwing at a prospect for dinner usually involves something novel: the target is not at the same distance or the same elevation as one of your standards; perhaps it is moving, too. And throwing, much more than such ballistic mo-

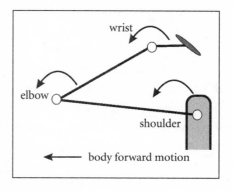

tions as spitting, involves a structured plan. Indeed, planning a throw has some nested stages, strongly reminiscent of syntax.

The highest velocity action is in the wrist movement, but planning it requires you to take account of what the elbow is doing: wrist flicks, where mistakes matter most, are nested inside elbow uncocking. What you want to achieve is a certain launch velocity, but you want the launch to occur at just the correct angle to the vertical. That's not a matter for the wrist alone. You need to take into account what the elbow is doing—or rather, since this is advance planning, what the elbow plan is. You have to estimate—guess, in other words—its motion.

Elbow planning needs to know what the shoulder is doing. And the shoulder too has a forward velocity due to what the whole trunk is doing, that forward velocity added by the legs. So planning a throw is a nested problem, just like understanding "I think I saw him leave to go home."

You don't do this as some assembly line during "get set,"

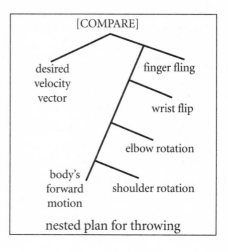

nested plan for throwing

marching from the fingers back up to shoulders, but you do have to juggle the finger, wrist, elbow, shoulder, and body plans, coordinating until you get the overall symphony right (as judged by your memories of similar-but-not-identical situations). While there are a number of combinations that might suffice for a given target location, they have to hang together in just the right way or you'll miss the target and go hungry. You need, in other words, a coherent plan: all of the parts (and there are about a hundred muscles involved) have to form an internally compatible plan. So, if the target is not standing at the location of one of your well-rehearsed set pieces, you need to make a novel, staged, coherent plan. And then execute it in an eighth of a second, getting all those muscles to come in at just the right time and with just the right strength.

It was probably an expensive bit of Darwinian engineering, to get throws to be as good as what even eight-year-olds can accomplish on the playground of my neighborhood school. Fortunately, once paid for by its usefulness in feeding the family with ever more frequent helpings of high calorie, nontoxic food, the neural machinery can perform other kinds of planning tasks as well, even for free.

E VEN without something for "free," this sounds like heresy. That's because arguments about useful adaptations seem to assume (even when there is no need) that one cortical area is dedicated to one function. As I mentioned earlier, you'd almost think that better throwing abilities ought to raise a bump on the head that could be labeled Hand-Arm Planning Center. It would be separate, of course (another beginners' error), from another bump labeled the Language Module. There are two quick arguments that nicely serve to illuminate this understandable beginners' mistake.

One is that there may not be sufficient genetic variety to expand just one little region of the brain; the only variety available may be to increase major portions of the brain together—and that's what most of the evolutionary data suggests. "Increase one area, increase them all" may be the general rule. There are important exceptions: some lineages can increase the olfactory areas of the brain without also increasing the cerebral cortex. In general, raising a specialized bump is not an available option, however efficient it might seem as a first guess. (As I'll mention later, our intuitive notions of biological engineering often do not correspond to Darwinian reality, not any better than our intuitive physics matches up with Newtonian reality—and certainly not Einstein's.)

Second, though some of the best-understood regions of the brain such as primary visual cortex seem rather dedicated to a specialty, much of association cortex seems multifunctional. Certainly there is much evidence suggesting that oral-facial movement planning can overlap with that for hand-arm—and with that for language, both sensory and motor aspects. So we might even see "improve one function, improve some others in passing."

WHATEVER economist said that "There is no such thing as a free lunch" obviously didn't absorb the lessons from Darwin and his successors. Pay via natural selection for one functionality like planning or language, and you may get the others such as music mostly "for free."

Let us assume that, however we got them, we have some brain circuits that are capable of running a process for making multistage coherent plans, and judging them for quality against your memory of what's reasonable and safe. Can you use them for other movement sequences as well as hand-arm?

Not only is there a great deal of multiple use in evolution but you can see a nice concrete reminder on many a street corner—some missing concrete. Wheelchair considerations paid for curb cuts but soon 99 percent of their use was for things that would never have paid their way—baby carriages, grocery carts, skateboards, wheeled suitcases, bicycles, and so on. Maybe one of those secondary uses will eventually pay for further improvements but pay-before-using is not required.

When did spare-time uses develop for the neural machinery for planning throws? Hammering was likely the earliest, though I wouldn't push cause and effect too far here. With shared machinery, you can have coevolution with synergies: better throwing might improve, in passing, the ability to hammer accurately. And vice versa. Though I am fond of accurate throwing as an early prime mover, remember that any of the uses of structured thought might improve the others, different ones at various times. Even before the transition, language probably started paying its way.

One of the free uses of the curb cut has already paid for a subsequent improvement. I can remember when traffic jams occurred at the wheelchair ramps at airports. The wheeled suitcases would queue up, awaiting their chance at the slot, and so in the newer airports, curb cuts were made as wide as the crosswalk. When robotic developments enable both wheelchairs and suitcases to climb stairs, they'll find curbs easy. Indeed, the curb cuts may become obsolete for their original functions, though still frequented by bicycles and skateboards. By the time that their original functions are forgotten, skateboarding will probably have evolved into a religion. The skateboarders will surely claim the curb cuts as their ancestral sites of worship—and try to exclude pedestrians.

In seeing the curb cut as created for its then-current best use, the skateboarders will be making the same inference that we make now when we posit that the evolution of the big brain is all about intelligence. Maybe. Maybe not.

Secondary use initially gets a free ride, and it doesn't necessarily retire the original use in the manner of Darwin's example, the fish's swim bladder that turns, after an intermediate period of dual use, into a lung. (That's where Darwin also cautions about going overboard on adaptations via natural selection, observing that conversions of function can also be quite important.)

The structure can remain multifunctional. The name often changes to reflect the most obvious high-order use—and since the brain is very good about multiple use, maybe our high-order uses ought to be seen in this curb-cut context. All of this, of course, is meant as parable: I want you to see the evolutionary development of complex thought as a parallel to the recent expansion of curb-cut uses.

But just because some secondary use is possible doesn't mean it instantly happens. We were likely capable of structured music (what you're doing when the left hand plays a different tune from the right hand) just after the big transition, long before Western music got around to using it about a thousand years ago. Just because novel symphonies of hand-arm movement commands had been getting better and better for a million years doesn't mean that secondary use for spoken language had to happen.

S o when do kids pick up structured stuff from their experiences in life? Modern children can do it from speech between 18 and 36 months of age, even before they can tie their shoes (fine movement control matures more slowly)—provided, of course, that their culture provides them with lots of examples of structured stuff to puzzle over.

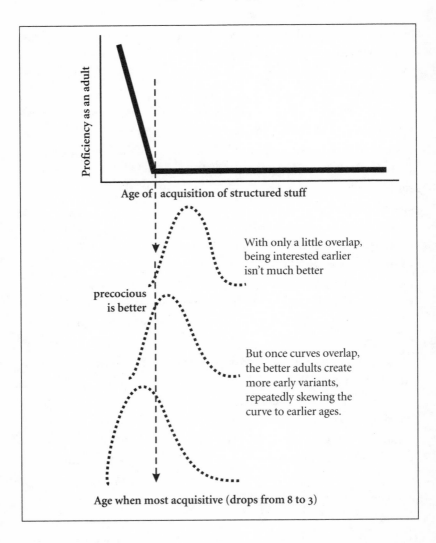

Let us say that, back 150,000 years ago, it was only when practicing accurate throwing at age eight that a lot of novel structured stuff was experienced. Protolanguage was perhaps around, but short sentences can be understood without looking for structural hints as to roles. All of that practice throwing at novel targets served, let us say, to softwire the brain in the manner of learning, so that adult performance was better on structured stuff.

But picking up structured stuff also depends on an individual's acquisitiveness, as in those modern kids picking up many new words a day. Some kids are acquisitive of structured stuff earlier than others; it's probably some bell-shaped curve. And their performance as adults on structured stuff depends strongly (if we are to judge from the deaf kids without sign language environments) on a window of opportunity (to be dramatic, let us say it opens at age two and flattens at five years of age, that thick line).

So being precocious pays off as an adult, because you do a better job of softwiring for structured stuff, having done it earlier than the average child. There are three cases to consider:

- The age when interest doesn't really overlap with the window of above-average softwiring opportunity. So the more precocious kids (left side of the top bell curve) aren't better as adults than the average kids, no matter when culture exposes them to structured stuff.
- The precocious kids are overlapping with the sensitive period for softwiring the brain for structured stuff—that earlier-is-better segment of the thick line—but culture doesn't provide any structured examples that early, so they never tune up when earlier is better. (The tragedy of the modern deaf kids with hearing parents, but also what anatomically moderns might have been like before syntax.)
- Culture provides early examples of structured stuff via, say, speech. Now earlier is better for eventual adult performance. And those successful adults are providing the next round of variations centered on their (skewed) average via Darwin's inheritance principle. So the more precocious of the offspring of the previously precocious are even better as adults. With this,

successive generations can keep marching to the left, back up
the earlier-is-better part of the thick line.

This is just another aspect of what are called epigenetic factors in
development, where the environment serves to trigger an alternative
path in development. In the case of plants, it is sunlight that provides a
cue to a new branch as to whether to grow upward and sprout leaves or
to grow downward and develop root hairs. Culture too can provide
important cues for making development choices and directional selection can move the succeeding generations up the curve. Evolution
interacts with development via the environment (the research area is
known as EvoDevo) and that's why nature-nurture and genes-culture
are such false dichotomies.

For now, note that none of this requires very much of what is usually posited by the archaeologists: some new genes to initiate the transition. Even the tweaks in acquisitiveness might come later, if the
curves originally overlapped the window of better softwiring opportunity. In just one generation of kids finally being exposed to structured
stuff at earlier ages, the next generation of adults would be far more capable, thanks to softwiring in youth. (In fact, it need not wait until the
next generation: older children are usually the frequent companions of
younger children, and so a mother speaking structured sentences can
infect her children, who themselves infect other children a few years
later.)

So the simplest version of a rapid ascent to structured thought looks
like this:

- PRE: Structured stuff is learned in later childhood with accurate throwing and toolmaking.
- TRANSITION: Some older children and adults manage to
  slowly add structure to protolanguage utterances.

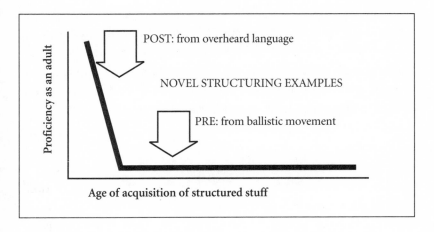

• POST: Young children are now exposed to structured stuff via the speech of caregivers, even before they can "tie their shoes," and they then softwire their brains to really fly as adults. But because structured stuff via one route may carry over to such things as planning and logic, most of the structured suite pops into place in a few generations.

So the archaeologist's summary of what "behaviorally modern" involves—abstract thought, planning in depth, innovation, symbols, storytelling—overlaps with higher intellectual function. But a "symbolic" formulation doesn't hint at the underlying unity that I've been covering here.

My point is that much of this behavioral modernity is structured—syntax, contingent plans, music, logical chains, narratives, games with rules, house-of-cards analogies—and that we compulsively guess to fill in missing pieces in the inferred structure. Guessing at structured stuff means we can make a lot of mistakes, so we have to be constantly concerned with quality.

Quality and coherence are also what limits creativity. I will devote an entire chapter to creativity after saying something about how much of higher intellectual function seems half-baked, what you ordinarily see in a prototype rather than a finished, well-engineered product. Perfection you don't get, not from Darwinian evolution. And the quality controls are spotty. But culture—especially education and medicine—can sometimes patch things up, if society works hard enough at it.

*A*s masters of illusion, specialists in the evolving art of social control, [shamans] held exalted positions. It could not be otherwise. As equals, they could never have done what they had to do, indoctrinate people for survival in groups, devise and implant the shared memories that would make for widening allegiances, common causes, communities solid enough to endure generation after generation. . . . The ceremonial life promoted, not inquiry, but unbending belief and obedience. . . . To obtain obedience it helps if shamans . . . can create a distance between themselves and the rest of the group. . . . One must appear and remain extraordinary (by no means an easy task when one is a long-time member of a small group), look different with the aid of masks . . . and sound different, using antique words and phrases, reminders of ancestors and a remote past, and special intonations conveying authority, fervor, inspiration.

—JOHN E. PFEIFFER, 1982

Abduction scene, Kolo Cave, Tanzania
(From copy at National Museum of Kenya, Nairobi)

9

From Africa to Everywhere

*Was the still-full-of-bugs prototype
what spread around the world?*

W ITH THE TRANSITION TO *Homo sapiens sapiens*, we probably saw ourselves in a new light. The premodern mind
might well have lacked our narrative tendencies—and,
with them, the tendency to see oneself as the narrator of a life story, always situated at a crossroads between the alternative interpretations of
the past and the paths projected into various possible futures.

Scientists are uncertain storytellers. Though we may stand atop a
stable pyramid of certainties, laboriously established by our predecessors, we are always attuned to the uncertainties of the scientific frontier—and so we are professionally uncertain. We try to bring coherence
out of chaos and we sometimes get it wrong. We expect to be telling a
somewhat different story, ten years from now.

T HE campfire storytellers who first attempted to reconstruct human
origins probably concentrated on the travels of their grandparents.

But at some point, armed with the abstract imagination that characterizes modern humans, they created "origin stories" of the Adam-and-Eve variety, embellished down through the generations, which attempt to account for human beginnings.

The Cherokee Indians tell of a creator who baked his human prototypes in an oven after molding them from dough. He fired three identical figures simultaneously. He took the first one out of the oven too early: it was sadly underdone, a pasty pale color. Creators may not do things perfectly the first time but in creation myths their actions are always irrevocable—so the pale human was not simply put back in the oven to cook a little longer. It remained half-baked.

But the creator's timing was perfect on taking the second figure out of the oven: richly browned, it pleased him greatly and he decided that it would become the ancestor of the Indians. He was so absorbed in admiring it that he forgot about the third figure in the oven until he smelled it burning. Throwing open the door, he found it was black. Sad, but nothing to be done about it. And such were the origins of the white, brown, and black races of mankind.

The Cherokees are not alone in being ethnocentric. There is sometimes an ethical lesson to the origin stories, improved over time, but whatever original facts were part of the origin story tend to become less accurate over time, with enough retellings. As the anthropologist and poet Loren Eiseley said, "Man without writing cannot long retain his history in his head. His intelligence permits him to grasp some kind of succession of generations; but without writing, the tale of the past rapidly degenerates into fumbling myth and fable."

SOME religions have learned not to "bet the farm" on the literal truthfulness of a particular creation myth—not to create a situation where truly valuable teachings become as tainted as the tooth fairy when one of the props turns out to have been oversimplified or imagined. Studies of human evolution, using such tools as changes in DNA, are now rapidly recovering some of the basic facts about who migrated where and when.

I won't attempt to explain the mechanics of the DNA dating method except to note that it is a lot like the linguists' technique of noting similar word roots in related languages. Whereas the English say "fist," "finger," and "five," the Dutch instead say "vuist," "vinger," and "vijf." The Germans use "faust," "finger," and "funf." That's because all three languages are descendants of the Germanic language family. Sir William Jones, while serving as a judge in India in 1786, noted that Sanskrit shared a number of features with Greek and Latin, suggesting an ancient language in common among Mediterranean peoples and those of the Indian subcontinent.

The family of Indo-European languages, as it is now called, represents the splits and separate evolution of dialects over the centuries. Finnish, Estonian, and Hungarian are not members of that language family, being more related to the Uralic languages in Asia, and testifying to a thousand years of invasions of Europe from the grasslands of Asia.

One can play the same game with genes that differ between groups, and even estimate the time at which one ancient tribe split off from another, likely because of some migration that caused them to lose touch with one another. For accuracy, it's best to focus on genes that are not shuffled with each generation, such as the Y chromosome which is passed only from father to son. The mitochondria inside each cell have their own genome, and it is passed only through your mother, who got it from her mother, and so on back. As such, they are more stable.

My mitochondrial genes, for example, are from Sweden, as that's where my maternal great-grandmother came from. Similarly, my Y chromosome is likely from England, though there's the possibility of one mutation since then (the average rate of mutations is what allows estimates of the time since two groups split). But, thanks to gene shuffling as sperm and eggs are formed, the rest of my genes are far more of an eclectic mix of hundreds of ancestors in just that two-century time frame.

The first migration out of Africa was *Homo erectus* about 1.7 million years ago. Some more likely followed, and the most recent major emigration from Africa is what populated the modern world with our species, *Homo sapiens sapiens*. The migrations are inferred from both conventional archaeological dating and the molecular clock.

I f you haven't been following the anthropologists' incessant dating and re-dating of important events, you might still think that the last Out of Africa was at 100,000 years ago, into Israel—and that behaviorally modern was much more recent, say when blade technologies and cave art appeared in Europe 30,000 to 40,000 years ago.

This was awkward, as it required the non-European groups to develop simultaneously their own behaviorally modern transitions, e.g., that the Australian aborigines did it on their own, and somehow at the same time as the new Europeans.

But at the turn of the twenty-first century, the best estimates for both dates changed—and they converged on about 50,000 years ago. Not only is Israel recognized as not being very exotic (much of the time, it has an African flora and fauna), but the mitochondrial DNA dating for Out of Africa—originally supporting the 100,000-year emergence—was redone using more appropriate portions of the mitochondrial genome. A period between 60,000 and 40,000 years ago became the more probable time for the immigration of *Homo sapiens* into

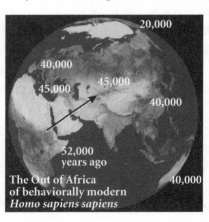

20,000

40,000

45,000   45,000

40,000

52,000
years ago

The Out of Africa
of behaviorally modern
Homo sapiens sapiens

40,000

the more exotic parts of Eurasia. The Y-chromosome gives similar bracketing dates. One estimate is that about 1,000 people—the size of a tribe and perhaps speaking a common language—emigrated into Eurasia and multiplied.

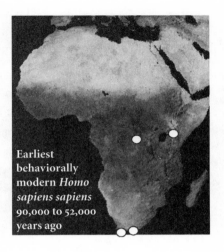

Earliest behaviorally modern *Homo sapiens sapiens* 90,000 to 52,000 years ago

Furthermore, various indicators of behavioral modernity have moved back in time (recall that figure on page 69). While what we'd like is a marker for syntax or long-range contingent planning, we must often rely on the behavioral B's: blades, beads, burials, bone toolmaking, and beautiful. Barbs on bone tools for fishing are seen in the Congo about 90,000 years ago. In the same period, on the southernmost coast of Africa, come the earliest signs of something "symbolic" in the form of cross-hatchings on red ochre. By 52,000 years ago in east Africa, beads can be seen. Clearly they were beginning to think something like we do.

So it now appears that humans were behaviorally modern before the last great Out of Africa. Many paleoanthropologists are now happy with talking about most behaviorally modern abilities emerging in Africa between 90,000 and 50,000 years ago—and that resolves the thorny multiregional problem for behavioral modernity, as one can assume that it developed once in Africa and then spread around the world. Certainly better language and planning would have made it easier to scout out new territories.

Almost all anthropologists (including the ones favoring the multiregional hypothesis for occasional interbreeding between Neanderthals and moderns) believe that all modern peoples on all continents are overwhelmingly descended from modern Africans. Between 60,000 and 40,000 years ago, behaviorally modern *Homo sapiens sapiens* spread into Asia, perhaps on both northern grasslands and southern shoreline routes.

I won't cover the many secondary migrations since then except to note that for the northern route into the steppes of central Asia, we have another important data point from Y chromosome studies: All non-African peoples (well, at least males) share genes that were present in central Asia about 40,000 years ago, suggesting a secondary center of spread from there that displaced the Neanderthals in Europe and the remaining *Homo erectus* in China. So think Out of Africa, followed by Out of the Steppes of Central Asia. First out of the cradle, then out of the nursery.

THIS history of the Out of Africa migrations makes the old concept of race look very simplistic. There are at least three things that contribute to our everyday concepts of race: teams, adaptations, and appearances.

As I mentioned earlier, we will form up teams, quite spontaneously. It's an important aspect of what makes us human. Stewart Brand wrote about being part of a spontaneous team of passers-by that formed up in the 1989 San Francisco earthquake to rescue people from collapsed buildings. But we often form opposing teams for trivial purposes, just as easily. "Us against them" is often founded on ethnicity or religion, but fans of one sports team may work cooperatively to harass another such ad hoc fan club. There are not major genetic differences between the people who fight one another in the Balkans or Northern Ireland. Even if they didn't form up teams around a shared religious background, they'd likely form up teams based on some other shared interest. But something that can be heard (a style of speech) or seen (different styles of dress or facial appearance) can easily bias which team you join. Ethnicity, based on such behavioral choices, is a significant part of the older notions of race.

But some differences are indeed biological. Appearances sometimes involve specific physiological adaptations, such as keeping the dust out of the eyes in Asia. In northern Europe, the adaptation involved making the most of scarce sunlight for vitamin D production in the unclothed patches of skin. The equatorial Maasai's long limbs are appropriate to losing heat. The short limbs and long body of the far-northern Saami and Inuit peoples are best for conserving heat. Some regional genetic differences produce more subtle effects: though there is no difference in the rate of having identical twins around the world, Africans give birth to a lot more fraternal twins than do Asians (Europeans are right in the middle). It will not be surprising if some *average* aspects of mind—say, styles of problem solving—turn out similarly to differ among regional groups, even if individuals in each group span the whole human range of variability.

But like having a "Roman nose," much of race is just being part of a large extended family and sharing the family resemblance. From appearances, it is easy to create African, Asian, and European clusters but there is nothing sacred about a continent—particularly Eurasia!—and things can obviously be subdivided from regional extended families all the way down to the characteristic appearance sometimes seen in a small village. Experienced travelers claim to recognize hundreds of categories. Some recognizable appearances come from recent mixes, as in the Mexican combination of recent European and ancient Asian genes.

European "whites" themselves have had a lot of recent gene contributions from the numerous Out of Asia episodes between when Attila's Huns crossed the Rhine in 436 to the great siege of Vienna in 1683. Then there was the African invasion of France in 732. African influences usually came north via the gradual spread of Mediterranean genes but also because of mass movements; after being forced out in 1492, many Spanish Jews emigrated to northern and eastern Europe. So Europeans are quite a mixture; the main thing we have in common biologically is a former need to make our skins unusually pale (a condition we often try to disguise via sun tans). Present-day African Americans average about one-quarter European genes and, however useless or pernicious that racial identification has become for most purposes, it occasionally is still an important clue to diagnosing an inherited disease such as sickle cell anemia.

That said, let us consider the consequences of spreading humans around the globe at a time not long after we first became behaviorally modern.

WHILE structured thinking may be one of the aspects of human uniqueness (and, when combined with ethical judgment ability, certainly a candidate for our crowning glory), I suspect that evolution hasn't yet tested it very well—that it is clunky and perhaps dangerous at times. Our emotional value judgments are far older and better tested than our intellects. Emotions are handy when decisions must be made quickly but they are also overly broad, lacking precision and nuance. Many problems arise when snap judgments substitute for deeper consideration.

Consider some of the consequences of acquiring higher intellectual functions:

- Our language instincts have become strong enough for orators to hold us spellbound, to urge us off to war.
- Our multistage planning is now good enough to prepare for a prolonged war, not just raids.
- Our compulsive search for coherence often results in finding hidden patterns where none exist. Astrology is one of the more innocuous examples (compared to religious cults, more later) of our tendency to find patterns in random noise.
- Our logic is so impressive (when it works) that we can convince ourselves that there are no other possibilities than what

we have eliminated with our logic—even though the history of both science and politics is full of examples where we were blindsided, or where our logic instead led us down a garden path into a mire.

Cowboys have a way of tying up a steer or bronco that fixes the brute so that it can neither move nor think, like the proverbial deer frozen in the headlights. This is the hogtie, and it is what our rationality sometimes does to us, freezing us when we ought to keep searching.

There are ways around this, as when we teach about the common logical fallacies. Just as medications can fix some problems, it is hoped that education about pitfalls might alleviate others. Teaching "critical thinking" skills in school is one way to combat the pervasive misleading information and logic that bombards us daily. We can learn to routinely ask: Why is this free? What are they really selling? What did they avoid mentioning? Do those statements really contradict one another? Important relative to what? Does that conclusion necessarily follow? Is the definition adequate? Is there ambiguity here? What's being assumed? Can you rely on the alleged authority? And when everything seems to hang together nicely, be wary of arguing in a circle—you may just be using different synonyms for the start and finish.

There is even an organized practice of trying to find people lacking critical thinking skills. The technical term is "trolling for suckers"—locating the fraction of the population that is truly impaired in their judgment and then attempting to part them from their money. "Sucker lists" collect the addresses of gullible people who have responded to something-for-nothing bait that is so improbable that most people would ignore it—but nibbling at it marks you for further attention. As in the New York joke about selling the Brooklyn Bridge, "If you can believe that, then I've got a bridge that I'd like to sell you."

Schemes like this require communicating the bait to a large number of people in order to find the unfortunate few. Nearly free email has now enabled mass mailings on a scale never seen before. The everyday lament, "How could anyone believe such stuff?", has a sad answer. Such ploys can pay off, sometimes not immediately but later on the followup (the free "bait" and then the "hook"). But we are all gullible on some occasions, often early or late in life, or on some subjects at all times.

This organized exploitation of intellectual shortcomings could be controlled by ethics and laws. But there are many everyday defects, shared by everyone to some extent, that make you wonder if *Homo sapiens sapiens* was really ready for prime time when everything expanded 50,000 years ago.

C omplex thought presumably underlies the entire suite of higher intellectual functions. We can operate at levels that are not easily translated into words, as when we mull over a puzzle. I'll have much more to say on the subject of levels of organization in the next-to-last chapter, but first let me mention some flaws in the more fundamental cognitive processes, where it looks as if we are still plagued by the crudeness that usually characterizes anything that is a "first of its kind":

> *Categorical perception* can put blinders on us, so that we cannot see the nuances. Japanese reared without hearing the English /L/ and /R/ sounds will lump them together and hear a Japanese phoneme that is in between in sound space; "rice" and "lice" sound the same to them. Newborn infants can hear the difference, but soon a category forms up around the sounds most often heard, and variants are conformed to the new standard. (It's known as the magnet effect or category capture.) Perception can also fill in missing information erroneously, as when blind spots in our visual world are filled in (look at wallpaper and, instead of seeing a featureless spot where your pho-

toreceptors are missing, you see the area filled in by the surrounding patterns).

We also do this fill-in over time, as when a light flash in one location is followed by a second light flash nearby—and we report seeing a smooth movement of the light, filling in all the intermediate points. A striking example occurs in viewing cave art with the aid of a flickering oil lamp. Between flashes, one's eye position drifts a little in the profound darkness and, when the next flicker again illuminates the scene, it looks to us as if the depicted animal smoothly moved!

While we share such perceptual inaccuracies with most primates, we have higher-order cognitive versions of category-capture and fill-in as well.

Our *memory mechanisms* are not very good at avoiding substitutions or keeping things in order. A child taking part in a collaborative project, when later asked who did what, will often think that she performed an action that the videotape shows was performed by another child.

I'm not referring here to what Henny Youngman said, "After you've heard two eyewitness accounts of an auto accident, you begin to worry about history." I'm talking about what happens much later, even if you get it right initially. Even when you initially succeed in recalling a sequence of events, you may make a mistake in recalling the event weeks later. If you scramble things once, it may have consequences a month after that, as if you had overwritten the correct memory sequence with your erroneous recall.

The memory expert Elizabeth Loftus likes to say that "Memory, like silly putty, is malleable. . . . The inaccurate memories can sometimes be as compelling and 'real' to the individual as an accurate memory." Keeping things in the right order is often important for structured thinking, and it looks as if evolution didn't get around to fixing the flaws in memory mechanisms.

Changing the name of something is, of course, a standard attempt to manipulate your memories, perhaps to run away from a problematic reputation. (Cynics would note that both my local telephone company and my bank have changed their names twice in recent memory.)

Our *structured judgment* may not be up to the task even when we structure our thoughts successfully, as in those fallacies of logic. And as merchants know all too well, our decision-making is easily swayed by the last thing we happen to hear. Psychology texts are full of examples about the unwarranted emphasis that is often given to some minor aspect.

Vivid examples can capture our minds and override other considerations. Although we might spend all day carefully considering the documented facts about frequency-of-repair records when shopping for a new car, our judgment is still notoriously easy to sway with just one nonrepresentative example. Someone at a dinner party complaining about repairs to their top-rated car is often sufficient to override our logical consideration of the average repair experience. We ought to treat the new example as just part of the range of variation that led to the average we researched. Instead, captured by the vivid example, we go out the next day and buy the second-choice car.

Any narrative provides an attractive framework, when competing with dry facts detached from stories. Ronald Reagan often took advantage of this when he was president of the United States, telling an easily appreciated story of some one person—and letting this carefully selected example serve as a rationale for a favored government policy. Vivid stories can be used to smother inconvenient facts.

Searching for coherence, we sometimes "find" patterns where none exist—as when imagining voices when it is only the sounds of the wind, or trying to force fit a simple explanation on a complicated set of relationships. "It all hangs together" is what makes for strong belief systems and allows all sorts of actions to be rationalized.

We offer *reasons,* often several deep, for an action or a belief. Some considerations, perhaps ethical ones, can override others. *Rationalizations* are untruthful inventions that are more acceptable to one's ego than the truth. We fall prey to logical fallacies; even snails assume "after this, therefore because of this" and you'd think that evolution could have kept us from falling for it so often. *Reasoning* often involves a chain of reasons, a considerable limitation because the reality is usually a more complex web of interacting causes.

There are *disconnects between thought and talk,* as in that blues lament of Mose Allison, about when "Your mind is on vacation but your mouth is working overtime."

Conditionals and pretense work surprisingly well, considering how little evidence there is for such abilities in the great apes. We have an ability to entertain propositions without necessarily believing them, distinguishing "John believes there is a Santa Claus" from "There is a Santa Claus."

But we aren't born that way. The ability to play a role in "doctor" or "tea party" arise later in the preschool years. Do we later remember what was pretense and what was real? Not always.

Source monitoring (tagging facts with where you learned them) often fades with time so that "facts" become detached from their supports. The day afterward, you may know it only happened in a dream—or that you only planned to say it but didn't actually utter the words—but will you lose that pretense tag in another month?

Concreteness is seen in a few modern people who answer very literally to any example of figurative speech, who are unable to rise beyond the most basic interpretation. But most of us are very good at backing off and treating a question more abstractly, looking for the metaphor. When someone starts to lose this ability, physicians suspect damage to the frontal lobes and go looking via diagnostic brain imaging.

R ATIONAL argument alone often cannot overcome those who simply and passionately believe. Yet logic is often bent and distorted in the service of those belief systems; it can even override everyday experience. As Fyodor Dostoevsky noted, "But man has such a predilection for systems and abstract deductions that he is ready to distort the truth intentionally, he is ready to deny the evidence of his senses in order to justify his logic."

I doubt that this was a problem before the behaviorally modern transition. There is nothing like logic in the aid of strong beliefs to provide the motivation to override ethics and find hypocritical excuses for committing acts of violence. It is most familiar from extreme political beliefs, but consider two examples of how professedly peaceful religious cults have, once they became wealthy via give-us-everything contributions from their members, turned to using biological and chemical terrorism.

In 1984, members of the religious cult of Bhagwan Shree Rajneesh sprayed the salad bars of four restaurants in The Dalles, Oregon, with a solution containing salmonella. The idea was to keep townspeople from voting in a critically contested local election; 751 people became ill. This cult merely obtained mail-order biological salmonella samples and cultured them. (This is low-tech kitchen stuff.)

The second cult, in contrast, recruited technically trained people in considerable numbers and engaged in indiscriminate slaughter. Aum Shinrikyo ("Aum" is a sacred syllable that is chanted in Hindu and Buddhist prayers; "Shinrikyo" means supreme truth) is a wealthy religious cult in Japan (recently renamed Aleph), with many members in Russia. Their recruiters aggressively targeted university communities, attracting disaffected students and experts in science and engineering with promises of spiritual enlightenment. Intimidation and murder of political opponents and their families occurred in 1989 by conventional means, but the group's knowledge and financial base allowed them to subsequently launch substantial coordinated chemical warfare attacks.

In 1994, they used sarin nerve gas to attack the judges of a court in central Japan who were about to hand down an unfavorable real-estate ruling concerning sect property; the attack killed seven people in a residential neighborhood. In 1995, packages containing this nerve gas were placed on five different trains in the Tokyo subway system that converged on an area housing many government ministries, killing 12 and injuring over 5,500 people.

During the investigations that followed, it turned out that members of Aum Shinrikyo had planned and executed ten attacks using chemical weapons and made seven attempts using such biological weapons as anthrax. They had produced enough sarin to kill an estimated 4.2 million people. Other chemical agents found in their arsenal had been used against both political enemies and dissident members. While they were also virulently anti-Jewish, apocalyptic scenarios dominated the sect's doctrine and the Tokyo attack was said to be an attempt to hasten the Shiva version of Armageddon. Only cult members would survive it, they claimed, thereby purifying the world by ridding it of nonbelievers. (No mention seems to have been made of the enormous cleanup job the true believers would then face.)

THE general problem here is the motivation provided by strong belief of all kinds and its narrow logic. As Dostoevsky observed, belief systems serve to distort new information, making it conform to preconceptions. It provides a narrowed focus, within which everything seems to hang together, and consequences follow from its logic. It reminds me of that aphorism: the person with one watch always knows what time it is. The person with two watches is always uncertain.

If you approach a problem from multiple directions at once, you learn to live with such uncertainty. Strong belief systems, however, often try to narrow you down to one source so that you cannot escape its logic. Educated people are not immune to getting trapped in such blinkered logic. Most technically trained people, including many physicians, do not have broad educations and they are often activists, trained to make decisions quickly and move on. The masterminds of modern terrorist movements are frequently personable, technically competent people from privileged or middle-class backgrounds, not at all fitting the usual depiction of their foot soldiers as the ignorant downtrodden. And such leaders, in the manner of modern executives everywhere, usually weed out the truly mentally ill as too unreliable. To get people to follow your orders, it helps if they can follow your logic.

Martyrdom is often the result of excessive gullibility, of ensnarement by narrowly focused logic. Some suicide bombers turn out to be educated people, trapped in the logic of some scheme where "everything hangs together." While the false-coherence problem is old, its consequences have escalated. "For the foreseeable future, smaller and smaller groups of intensely motivated people will have the ability to kill larger and larger numbers of people," Robert Wright writes.

"They'll just have to be reasonably intelligent, modestly well-funded, and really pissed off." Or really trapped by a compelling logic that reframes their existence.

Recall the Fermi paradox about extraterrestrial intelligence: "If intelligence is common in the universe, why haven't we seen them already?" (since some surely evolved technological civilizations before we did). This suggests that prior technological enlightenments elsewhere might have flickered only briefly before self-destructing.

Sometimes the strong belief and its logic is of religious origin ("God gave us this land"), sometimes secular ("All power to the people"). But all this has little to do with mental illness (that I will discuss in the last chapter), however great our reflex tendency to label some of the acts as "crazy." It has everything to do with our half-baked rationality.

DESPITE its virtues, anything that is "the first of its kind" tends to be awkward at first, rough around the edges. There is a growing suspicion that maybe modern humans are like that. Our intellects are a big step up but they appeared very recently in the ice ages, long after the human brain stopped enlarging. They're not well tested yet and are still prone to malfunctions.

However impressive our average intellect may be when compared with the other apes, remember that biological evolution often produces overblown features with major drawbacks. Peacocks have tails that hamper escape from predators. Elk grow antlers so wide they can no longer run through forests to escape wolves. Overgrown intellects have similar problems. An intellect with great persuasion and planning abilities sometimes produces dramatic results as the leader of a suicide cult.

As Desmond Morris once said, we prefer to think of ourselves as fallen angels, not risen apes. At least, we hope, evolution is still improving us. Alas, biological evolution doesn't perfect things—it just moves on to new "products" with a different set of bugs. (Sound familiar? And how often does your computer still hang or crash? We might be like that, not ready for prime time.) Even when we avoid hanging up from obsessions or crashing from epileptic seizures, we stumble over numerous cognitive pitfalls (usually without noticing).

Once you also recognize that we're *recently* risen apes, you realize that there simply hasn't been much time in which to evolve a less buggy version 2.0. Clearly, human cultural innovation is now in charge of getting the bugs out, not biological evolution. And we haven't made much progress yet.

We've arranged a global civilization in which most crucial elements—transportation, communications, and all other industries; agriculture, medicine, education, entertainment, protecting the environment; and even the key democratic institution of voting—profoundly depend on science and technology. We have also arranged things so that almost no one understands science and technology. This is a prescription for disaster. We might get away with it for a while, but sooner or later this combustible mixture of ignorance and power is going to blow up in our faces.

—CARL SAGAN, 1996

The way we think in dreams is also the way we think when we are awake, all of these images occurring simultaneously, images opening up new images, charging and recharging, until we have a whole new field of image, an electric field pulsing and blazing and taking on the exact character of a migraine aura. . . . Usually we sedate ourselves to keep the clatter down. . . . I don't necessarily mean with drugs, not at all. Work is a sedative. The love of children can be a sedative. . . . Another way we keep the clatter down is by trying to make it coherent, trying to give it the same dramatic shape we give to our dreams; in other words by making up stories. All of us make up stories. Some of us, if we are writers, write these stories down, concentrate on them, worry them, revise them, throw them away and retrieve them and revise them again, focus on them all our attention, all of our emotion, render them into objects.

—JOAN DIDION, 1979

10

How Creativity Manages the Mixups

*Higher intellectual function
and the search for coherence*

IMAGINE A TIME WHEN innovation was often nonsensical or even dangerous, because the brain couldn't improve the coherence and quality of a novel course of action during "get set." While perhaps you could produce juxtapositions of "look" and "leap," you couldn't improve their quality and so might leap before you looked. You'd be conservative, mistrusting innovation for good reasons. You'd stick to slow groping, testing each stage along the way rather than being able to make multistage plans that usually worked the first time out.

When Alexander Pope wrote about a little learning being a dangerous thing, and the need to drink deeply, he could have been writing as well about the good-enough thresholds for innovation. The beginner's mistakes could have made innovation have a negative payoff for a long time.

Still, creativity is a good thing, right? It *ought* to pay its way easily. We make the same assumption about intelligence, and the record of the past raises the same two cautions.

First, as Ian Tattersall said in *Becoming Human* when summarizing the archaeology, the behaviorally modern period "stands in dramatic contrast to the relative monotony of human evolution throughout the five million years that preceded it. For prior to the Cro-Magnons, innovation was . . . sporadic at best."

Second, if creative brains are such a good thing, why aren't there more of them? Most animals get by perfectly well without being innovative off-line: it's called "fumble and find." If they have to do something novel, where they don't have a stored movement plan to call up, they just muddle through, slowly feeling their way. (This is in contrast with the off-line "think first," doing most of the innovation in your head before making your initial move.)

If you have time to grope around, a goal plus some feedback along the way suffices nicely—and it mostly obviates the big problem with doing something for the first time, that of it possibly being dangerous. "Feel your way" may be grossly inefficient but it has a lot of virtues and most animals stick to it when doing something they haven't done before.

There are only a few things before the transition where muddle-through might not suffice. The ballistic movements are particularly interesting because feedback is too slow. The last eighth of a second of a throw cannot be corrected because it takes about that long for sensations from the limb to travel back into the spinal cord, decisions to be made, and new movement commands to travel back out to modify muscle activation patterns. A dart throw lasts only one-eighth of a second from start to launch, so you have to make the perfect plan during "get set." For a set piece target distance as in darts, you just call up with right commands from memory. But if it is a novel task, you have to think first and get it right before starting the throw. That kind of movement-command creativity had a payoff for at least two million years prior to the transition.

Since the distinction of anatomically modern from behaviorally modern humans was made, the usual explanation of the transition has been that language arrived on the scene and changed everything, including creativity. Yet language itself is just another example of creativity, once you move beyond stock phrases and start to speak sentences that you've never spoken before. While language surely makes it easier to spread around the results of creativity and build atop what others have tested, what we really want is the source of both language and ballistic movement creativity.

Here I will consider the demands that off-line creativity places on brain circuitry—and how "think first" creativity might have improved without any concomitant increase in brain size.

S TRUCTURED stuff is all very nice, and it has some payoff for set pieces like well-practiced throws. But that's learning, with creativity only at the beginning when fumbling around will suffice to explore the combinations. Coherence arises as you slowly get your act together. Learning is what keeps it together. Mammals play a lot while young, and much of play is good practice for making a living or discouraging competitors when adult. But all of that is easily done, to judge from other mammals, and likely it didn't require changes in brain size and organization for young hominids to learn in a similar way.

I'll restrict myself to creativity that is a "first of its kind" amalgamation, that involves stages, and with everything done off-line in the think first manner. Finally, it all needs coherence—to hang together so well that it usually works the first time out. It is that combination of requirements that makes it so difficult for the brain—yet children somehow manage it pretty well in the preschool years without being taught.

Creativity involving stages that is done *on the fly* but *off-line* during "get set"—that four-part combination is what evolution produced somehow. Eating meat regularly was an ancient payoff for it, even if it is no longer an everyday application for most of us. I am fond of the evolutionary lessons of throwing for a number of reasons that you may not share, but just try to think of another task at which humans excel that requires all those things. Now narrow down your list to those that have a big payoff at some point. Narrow that list down to those with repeated payoffs, where each time you improve your technique, you eat even more high-quality food, chase off even more predators, or get even more mating opportunities.

Also, I want to talk about the underpinnings of language and, while I have to use words to do it, I want simple mechanistic examples of antecedents—rather than the usual attempt to explain structured language in terms of its own obvious usefulness, once you have it. Even if throwing were only another one of the mostly for-free secondary uses rather than a prime mover, I'd still pick it for my teaching example for all the reasons I mentioned when discussing structured stuff. Throwing is structured, as in those nested commands for flipping the wrist while uncocking the elbow as you rotate the shoulder and lurch the body forward. And, outside of set pieces and slow-motion bowling, it often requires a novel set of commands, assembled on the fly as you "get set."

"The combinatorial engine underlying our number and language systems allows for a finite number of elements to be recombined into an infinite variety of expressions," said Marc Hauser in *Wild Minds.* But judging the coherence of a lumped-together assembly of concepts or movements has got to be very important. And, since you will rarely be right when you first start to get set, you need a bootstrapping procedure for improving the assembly off-line until it is good enough to act on. There may be various ways to do this, but the known process that can achieve coherent results from incoherent raw materials is the impressive one that Charles Darwin (and later Alfred Russel Wallace) discovered.

Darwin's quality bootstrap is not only seen at work on the millennial time scale of species evolution but also on the days-to-weeks time scale of the immune response. Can brain circuitry run a version of it on the time scale of thought and action, shaping up a good-enough movement program for ballistic movements? And perhaps other creative sequences, such as novel sentences? All in milliseconds to seconds?

ONE can summarize Darwin's bootstrapping process in various ways. A century ago, Wallace emphasized variation, selection, and inheritance. (It reminds me of a three-legged stool: evolution takes all of them to stand up.) But as I explain at more length in *A Brain for All Seasons* (from which the next two pages are adapted), there are some hidden biological assumptions in that three-part summary.

When trying to make Wallace's list a little more abstract to encompass nongenetic possibilities like cognitive bootstrapping, I listed six ingredients that seem essential to turn the crank (in the sense that if you're missing any one of them, you're not likely to see much progress):

1. There's a pattern of some sort (a string of DNA bases called a gene is the most familiar such pattern, though a cultural meme—ideas, tunes—may also do nicely).
2. Copies can be made of this pattern (indeed the minimal pattern that can be semifaithfully copied tends to define the pattern of interest).
3. Variations occur, typically from copying errors and recombinations.
4. A population of one variant competes with a population of another variant for occupation of a space (bluegrass competing against crabgrass for space in my backyard is an example of a copying competition).
5. There is a multifaceted environment that makes one pattern's population able to occupy a higher fraction of the space than the other (for grass, it's how often you water it, trim it, fertilize

it, freeze it, and walk on it). This is the "natural selection" aspect for which Darwin named his theory, but it's only one of six essential ingredients.

6. And finally, the next round of variations are centered on the patterns that proved somewhat more successful in the prior copying competition. *So variation isn't truly random; the starting place really does matter. And the next generation's starting place can, with success, shift a little.*

Try leaving one of these out, and your quality improvement lasts only for the current generation—or it wanders aimlessly, only weakly directed by natural selection.

Many processes loosely called "Darwinian" have only a few of these essentials, as in the selective survival of some neural connections in the brain during development (a third of cortical connections are edited out during childhood). Yes, there is natural selection producing a useful pattern—but there are no copies, no populations competing, and there is no inheritance principle to promote "progress" over the generations. Half a loaf is better than none, but this is one of these committees that doesn't "get up and fly" unless all the members are present.

And it flies even faster with a few optional catalysts. There are some things that, while they aren't essential in the same way, affect the rate at which evolutionary change can occur. There are at least five things that speed up evolution (and here I'll have to use species evolution examples; just remember that they can be translated into neural circuit equivalents).

First is speciation, where a population becomes resistant to successful breeding with its parent population and thus preserves its new adaptations from being diluted by unimproved immigrants. The crank now has a ratchet to minimize backsliding.

Then there is sex (systematic means of creating variety by shuffling and recombination—don't leave variations to chance!).

Splitting a population up into islands (that temporarily promote inbreeding and limit competition from outsiders) can do wonders.

Another prominent speedup is when you have empty niches to refill (where competition is temporarily suspended and the resources so rich that even oddities get a chance to grow up and reproduce).

Climate fluctuations, whatever they may do via culling, also promote both island formation and empty niches quite vigorously on occasion, and so may temporarily speed up the pace of evolution.

Some optional elements slow down evolution: "grooves" develop, ruts from which variations cannot effectively escape without causing fatal errors in development. And the milder variations simply backslide via dilution, so the species average doesn't drift much. Similar stabilization is perhaps what has happened with "living fossil" species that remain largely unchanged for extremely long periods such as horseshoe "crabs."

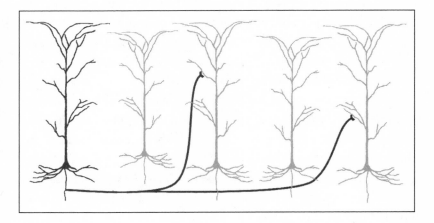

Are there brain circuits capable of running this sort of Darwinian process on the time scale of thought and action—say, milliseconds to minutes? Do they have enough of the speedup catalysts to operate so quickly?

That was the topic of my 1996 book, *The Cerebral Code*, and the answer appears to be yes for the recurrent excitatory circuits of the superficial layers of neocortex. They have patterned connections rather like express trains that skip a lot of intermediate stops. This should allow them to make clones of the spatiotemporal firing patterns, the codes that are used to represent a concept. Concepts can compete for ascendancy by running a pattern-copying competition biased by a virtual environment of feelings, drives, and memories. This can be as Darwinian as the copying competition between the bluegrass and the crabgrass in my back yard—except that what competes are codes for concepts and plans of action.

The circuits that seem capable of copying spatiotemporal firing patterns are found in most areas of neocortex, not just in the areas that might be involved in ballistic movement planning or language. But we still don't know how much of the neocortex makes use of this ability and when. Once, during a period of fetal or infant tune up? All the time, in all areas of neocortex? Where and when await experimental evidence, and the more interesting question—how to make subroutines out of the successful plans, so as to avoid running slowly through the whole Darwinian copying competition on subsequent occasions—hasn't been elucidated at all.

But let us assume that some brain circuits are capable of running such a process for making multistage coherent plans, and judging them for quality against your memory of what's reasonable and safe, biased by your emotions, drives, hopes, and fears. That gives us a prime candidate for the transition: the secondary use of ballistic movement planning circuits for the novel structured tasks of higher intellectual function.

Now let me briefly address Steven Mithen's 1996 notion of connecting compartments as a source of the mind's big bang. He proposes in *The Prehistory of the Mind* that before the transition, there were three different brain modules in the human brain that were specialized for "social or Machiavellian intelligence," for "mechanical intelligence" as in tool use, and for "natural history" as in a propensity to classify objects. These three modules remained isolated, suggests Mithen, from each other but around 50,000 years ago, some "genetic change in the brain" allowed them to communicate more effectively with each other, resulting in the enormous flexibility and versatility of human consciousness.

I find Mithen's notion appealing. It works well with my neural-circuits theory for how long-distance cortical interconnections might improve dramatically (and without any further brain enlargement or reorganization, at that). In the second half of my book that appeared that same year, *The Cerebral Code*, I explore the subject of rapid communications between distant cortical areas. The problem, from a brain theoretical point of view, is how to do it on the fly, without a slow learning procedure for each novel combination.

I suggest that the improved interconnections between areas occur when the distorting corticocortical interconnections (lots of jumble and blur from the anatomy) finally become temporarily restored via a physiological workaround involving sufficient redundancy. (It's something like an error-correcting code.) With such a plainchant chorus of sufficient size, you can recover a novel spatiotemporal firing pattern at the other end and pass it on (see chapter 7 of *Cerebral Code*) to a third cortical area, also unaltered. That ability to handle novelty routinely, dependent on finally reaching a critical mass that allows recovery of the original pattern from the blur and jumble, might well have contributed to the aforementioned transition.

So that's yet another transition candidate: you could have had structured stuff that was novel with off-line planning, but it was just too slow and too dependent on slowly learned code equivalents—until highly efficient common codes finally developed to allow concepts to be passed around easily between cortical areas over the distortion-corrected pathways. Even on the fly.

Our kind of high-order creativity—an ability to speculate, to shape up quality by bootstrapping from crude beginnings, yet without necessarily acting in the real world—is a recent thing, arriving well after the big brain itself. Toolmaking creativity doesn't look like the big actor in the ascent of humans—despite our usual notion of a versatile creativity being so important to being human. We are forced to consider our off-line creativity as a late, perhaps fortuitous development after other, more important, things were finally in place.

Of course, judging from the aforementioned malfunctions in structured stuff, you'd think that confusions were always a bad thing. But remember that a little confusion helps you to escape popping into standard answers. Being forced to pause, and sort out the possibilities for your confusion, can actually be helpful to creativity even though it slows down rapid decision making.

Intelligent people can juggle a half-dozen concepts simultaneously and make good decisions rapidly—and many of them seldom have a creative moment. They are so good at the standard answers and so eager to move on to the next decision that they never play around with nonstandard possibilities. (Physicians often get caught in this "keeping busy" mental trap, moving on rather than contemplating.) There is such a thing as being "too good" because, in much of life, there are no correct answers. You have to invent new ones and contemplate them for some time.

IQ tests tend to weigh heavily the speed of decision making and the number of concepts that you can juggle at the same time. Measuring creativity with standardized pencil-and-paper tests is difficult to do, but creativity and versatility tend to be at the heart of our everyday impressions about whether someone is particularly intelligent. Handling novel combinations and judging the coherence, redoing it to get a better innovation—that's what intelligence usually involves. It's why I implied, at the beginning, that beyond-the-apes intelligence and creativity both had the same stumbling block: judging off-line whether it all hangs together well enough to act on.

Still, it needs to be said that the light of evolution is just that—a means of seeing better. It is not a description of all things human, nor is it a clear prediction of what will happen next.

—MELVIN KONNER, 2001

From the origin of human language 100,000 years ago until the invention of writing 5,000 years ago, the oral tradition had been the principal creator, conserver, and communicator of human knowledge. Our brains are biologically adapted to the tempo of oral interaction in real time.

—STEVEN HARNAD, 2003

But surpassing all stupendous inventions, what sublimity of mind was his who dreamed of finding means to communicate his deepest thoughts to any other person, though distant by mighty intervals of place and time! Of talking with those who are in India; of speaking to those who are not yet born and will not be born for a thousand or ten thousand years; and with what facility, by the different arrangements of twenty characters upon a page!

—GALILEO GALILEI, 1632

Education is a technology that tries to make up for what the human mind is innately bad at. Children don't have to go to school to learn to walk, talk, recognize objects, or remember the personalities of their friends, even though these tasks are much harder than reading, adding, or remembering dates in history. They do have to go to school to learn written language, arithmetic, and science, because those bodies of knowledge and skill were invented too recently for any species-wide knack for them to have evolved.

—STEVEN PINKER, 2002

A wise man learns from his experience;
a wiser man learns from the experience of others.

—CONFUCIUS

11

Civilizing Ourselves

From planting to writing to mind medicine

W E LIVE AND LEARN and pass it on. By itself, that's a big improvement over what the great apes, lacking true teaching, can do—but with written language, passing it on works a lot better. We can even learn from authors long since deceased. Once agriculture allowed towns and specialized occupations to develop by 6,000 years ago, writing developed from tax accounting about 5,200 years ago in Sumer (the last six seconds of the movie).

With knowledge better able to accumulate and be taught, philosophy and the beginnings of science developed around 2,500 years ago in both Greece and China. That's also the time of the earliest historians such as Thucydides, who not only collected accounts of events but contrasted different reports of the same battle. Origin myths are not very good about doing that, though sometimes an alternate version was preserved if both were sufficiently poetic—as when the leadoff seven-days story is followed at Genesis 2:5 by the "second genesis" where God fashions man from the soil. It wasn't until the Darwinian revolution of the past century that science finally began to make some sense out of where humans come from.

Cultural evolution has now become much faster and more profound than what business-as-usual Darwinian processes are currently doing in biological evolution. Culture interacts with developmental processes, as in my example of how structured stuff could have become possible at earlier and earlier ages. A number of present-day human abilities have some potential for future elaboration even without natural selection helping. Reading and writing will serve as a good example.

The percentage of Europeans who were literate did not begin to increase substantially until the Protestant Reformation, with its emphasis on everyone (even women) reading the Bible themselves rather than just relying on a priest to interpret it for them. So until a few centuries ago, only a small percentage of the population had much opportunity to be improved by natural selection for reading abilities.

Still, at least 85 percent of us can be taught to read without much difficulty, even though we don't pick it up spontaneously in the manner of spoken language. It suggests that we surely had the latent capacity to read before 5,000 years ago—but that there need not be a "reading instinct" with its own brain area.

But there is a specialized brain area for reading. How can that be?

IN an adult human who can read, you can sometimes find an area in the brain that is essential for reading. It can do other things as well, but you can have strokes in small areas where the only obvious symptom is that the person loses the ability to read, without losing the ability to talk or write. Let me give you an example of what happened to my father, adapted from my account in *Conversations with Neil's Brain*.

One day he had a bad headache, quite unlike any other he ever had before. The next morning, he felt somewhat better and fixed himself breakfast, then went out to pick up the newspaper off the sidewalk.

Upon sitting down to breakfast and unfolding the newspaper, he discovered to his astonishment that he could not read it. The words weren't blurred. He could name the letters, but couldn't read the words.

Some days later, the neurologist asked him to write out a paragraph in longhand. My father accurately took the dictation. Asked to read aloud from his own handwriting, he couldn't. The neurologist had seen this before, but I was astonished that there could be such a "disconnect" between the two abilities. My father could name the letters correctly. He could often correctly guess at the shortest two- and three-letter words. But he often made errors when he would try to piece together longer words. His spoken language was normal and he understood everything that was said to him. He didn't have any abnormal blind spots that might interfere with reading and he could drive a car without problems. He just couldn't read anymore. A year later, he had recovered his abilities to read the newspaper, but he tired easily and wouldn't read for more than 20 minutes at a time.

So my father, like the other patients suffering from alexia without agraphia, appeared to have a specialized cortical area that was essential for reading. But where could such a thing come from in evolution, if reading is such a recent invention and the literacy rate was too low to expose reading to natural selection for its usefulness? One hint is that the strokes that cause such reading-only problems are in various places in various patients, not in some standard place common to all humans in the manner of, say, primary visual cortex in the back of the brain.

EXPERIENCE can rewire the brain and there is some evidence that reading abilities are wired up on the fly during childhood—as we say, they are "softwired" during development rather than hardwired in the manner of instincts. Self-organization from experience can create specialized areas of expertise in human cerebral cortex, when done early enough in life—and it changes the foundation on which later things can build. It's not just that the earlier you do it, the better as an adult, but that the order in which you learn things might matter.

The science of this is not well worked out yet, but let me give you an illustrative example from the research of my neurosurgical colleague George Ojemann. In studying the physiological organization of the left middle temporal gyrus (it's just above the left ear), he showed a strong association between the location of specialized reading and naming sites and, of all things, that patient's verbal IQ. He found that patients whose sites for reading were in the superior temporal gyrus, with naming in the middle temporal gyrus, had high verbal IQs. And he found the reverse in the patients with low verbal IQs. So some "residential layouts" of these functions in the cortex are more favorable than others. Why? Might early acquisition of reading skills lead to higher verbal IQ, simply because phonetic languages rely on writing to represent sounds, and the closer the reading areas are to auditory cortex, the better?

Tune in next year—parents are always experimenting on their children and researchers increasingly have the tools to image the brain's functional specializations and correlate them with test scores and learning history—but for present purposes, I just want this to serve as an example of how cultural feedback can cause self-organization, soft-wiring the developing brain to determine adult human capacities. It's a softwiring example, on the developmental time scale, of the hardwiring for acquisitiveness of structured stuff that I sketched out earlier on the evolutionary tweaking time scale: how overheard examples of structured stuff could allow softwiring for structured stuff at an age where it "takes better."

Culture works so well in the case of reading because we have childhood education for reading at a time when the brain is very plastic. Does softwiring add atop more instinctive stuff—say, sharing tendencies—to make a new type of person? Do professional musicians develop cortical specializations for harmony that the rest of us adults don't have? What about logical abilities, or being able to feel empathy well enough to routinely practice ethical behaviors?

We know that education matters, in the sense that ignorance is often expensive, but does properly staged education hold the potential for re-organizing the brain in profound ways? Even without genetic changes, the future baby might still be like modern babies when fresh out of the womb but become profoundly different in mental organization before reaching adulthood.

E DUCATION isn't required for some things. Children will learn to walk and talk without assistance, though swimming, reading, and writing usually require teaching. Recognizing someone from their face or the way they walk is a very difficult task, judging from many decades of attempts in artificial intelligence, but kids do them without being taught how.

It looks as if our minds come with good intuitions about some things, but not others. There is an intuitive physics about how objects fall and bounce but it's not very good. If you are out running and your keys drop out of your hand as you jump over a puddle, and if you man-age to take two more steps before you hear them hit the ground, where did they land—back in the puddle or after the puddle? Intuition will tell you to look back in the puddle, where you were at the time they dropped. (This is called Aristotelian physics—Aristotle failed to do the experiment that would have shown him that his impetus-based reason-ing was wrong. You have to study Newtonian physics to realize that the keys continue traveling forward at the same velocity as you were run-ning and so will land near your current position.)

We have an intuitive biology as well, in that we ascribe to something a hidden essence that drives its growth and makes it what it is. We often distinguish *how* from *why* using an intuitive notion of evolution that ascribes purpose to living things, as if they were designed for some goal or role. (And so we assume a designer, unless we have absorbed the lessons of the Darwinian process, those six essentials that can create very improbable beings after enough generations.)

We have an intuitive psychology that helps us understand other people, at least if we are old enough to pass that "theory of mind" test. We treat living beings as if they were animated by desires and beliefs. For people and pets, this works pretty well, though some people talk to plants too. And many of us, not just the cartoonists, intuitively ascribe higher intellectual functions to animals without language.

There is even an intuitive economics based on reciprocity that leads us to mentally keep track of who owes what to whom—and to punish cheaters and freeloaders, even at cost to ourselves. We have an intuitive notion of probability, though it is easily fooled. We have an elementary sense of number, though real arithmetic has to be taught.

This core of intuitions, except for the higher intellectual functions, was likely around even before *Homo sapiens* evolved. Add some protostructure such as framing and the closed-class words and you have a considerable advance over ape-level minds. Add some unstructured protolanguage and it becomes even more interesting, with gossip from well-known sources providing much experience at one remove. Hearsay was likely accepted without much questioning or reasoning about it (it often still is). Then structured thought itself finally emerges and we get additional ways of organizing knowledge and discounting the fallacious flotsam.

ADDING structured stuff atop the core intuitions took our ancestors into a realm of creativity where we got used to doing novel things. We got used to deciding quickly without much contemplation—and so we used those core intuitions quite a lot, even though evolution probably didn't get much of a chance to debug the combination of intuition and structured stuff.

While some of our core intuitions carry over to help us with modern economics, there are many areas (math, science, technology) in which we are simply unprepared for the modern world. "It's not just that we have to go to school or read books to learn these subjects," Steven Pinker writes. "It's that we have no mental tools to grasp them intuitively. We depend on analogies that press an old mental faculty into service, or on jerry-built mental contraptions that wire together bits and pieces of other faculties. Understanding in these domains is likely to be uneven, shallow, and contaminated by primitive intuitions."

Key concepts of, say, quantum mechanics or neuroscience actually require unlearning a lot of your intuitions. Neither Aristotelian nor Newtonian physics will help you grasp the wave vs. particle demonstrations. And always ascribing an actor to every action (as in those mandatory roles for verbs in the argument structure version of syntax) will get you into trouble searching for the seat of the soul or the center of consciousness—when what you need are concepts of distributed, self-organizing systems and how they handle novel inputs on the fly. (And create new levels of organization with just the right amounts of abstraction and anchoring—more in a minute.)

Perhaps premodern people simply weren't conscious in our modern sense, lacking most of that speculative "train of consciousness" that William James talked about.

I am conscious (aware might be the better word) of the chair supporting me as I read. I notice the sunlight from the window behind me. It is quickly fading and a blast of wind is rattling the trees near the house; another shower is likely to start pounding on the window. I am contemplating the cat, and she is contemplating the fire in the fireplace. She has one ear cocked back, a sure sign of an approach-avoidance conflict. I try thinking about a feline-based metaphor for a lecture but then become self-conscious, considering what the audience would think of me if I said something that silly. The music in the other room just switched from Brahms to some aborigine music. I am reminded of the Australian trip and start speculating about the prospects for a trip to the Galapagos. Which reminds me that I'd have to get some dental work done before going.

The here and now, the past and future, worries and delights, self and others—all are conscious aspects of mind. Sometimes an agenda intrudes, as I remember to check my watch, so that I leave in time for a pending appointment. There are nagging responsibilities that I push onto the back burner for now. I contemplate getting up from my comfortable chair to exercise but soon the moment passes. My mind drifts until consciousness wanes and lapses. The telephone awakens me and I quickly check my watch before I get up, heart pounding. A close call.

Consciousness is, however, more than just the minimum requirements of awake and aware. Many would emphasize that sensation becomes conscious only when it undergoes some further processing in the brain, as when it encounters past associations or becomes part of a plan for action.

Some of such consciousness is likely shared with the great apes, but probably not the speculative aspects. Nor are they likely to experience the pangs of conscience. That takes foresight, what augments the minor amounts of moral sense that the other apes have.

A central aspect of consciousness is the ability to look ahead, the capability we call "foresight." It is the ability to plan, and in social terms to outline a scenario of what is likely going to happen, or what might happen, in social interactions that have not yet taken place. . . . It is a system whereby we improve our chances of doing those things that will represent our own best interests. . . . I suggest that "free will" is our apparent ability to choose and act upon whichever of those [scenarios] seem most useful or appropriate, and our insistence upon the idea that such choices are our own.

—RICHARD D. ALEXANDER, 1979

O NE of the reasons that paleoanthropologists like to talk about consciousness at this point is that modern humans seem so much more capable of high-end cognitive functions. An umbrella-coverage term, something like the living-nonliving distinction, is attractive to many people. Personally, though I concede that we are conscious in ways that great apes are not, I often avoid the C word and other such big, loaded words when I am trying to describe things that I think of in more textured, fine-grain terms.

There's a famous quip by Francis Crick, about the border between the living and the nonliving forms of matter, something that used to cause a lot of debate in the first half of the last century. Well, said Crick, note that this all-important boundary gradually disappeared into just so much molecular biology. And, he suggested, the same thing was going to happen to consciousness as a concept—that it would disappear into just so much neurobiology. (I would add: We'll still talk about it, just as "alive" has remained a useful concept, but consciousness won't be a "thing" anymore.)

M EDICINE now calms the voices and delusions, dampens the obsessions and compulsions, and lifts the depressions. Besides patching us up, so we aren't hobbled, might it eventually "improve" us so that we perform in extraordinary ways?

I'm not a big fan of optimal performance. It usually sacrifices versatility and creativity, and so optimization often leaves you stranded, high and dry, when the time comes to move on to a job promotion or the next phase of life. But I can imagine much more useful cocktails of the everyday stimulants that will help students study better, help artists be more creative, help with multitasking. Espresso stands will specialize in them, customizing the mix for your desired mental set.

Most of these innovations, of course, will prove to be illusory, mere placebos. Some will prove dangerous even in mild daily doses, in the manner of nicotine and LSD. Others will be efficacious but will be overused. "If some is good, more must be better" is a fallacy recognized even in ancient Greek medicine (almost any medicine becomes a poison in sufficient quantities). For food and water, we fortunately satiate. But satiety often doesn't kick in for more novel substances such as tobacco or alcohol. We can indulge to the point of significant impairment, with enough softwiring changes in some individuals' brains so that backing up becomes very difficult (it is called addiction).

So consumer experimentation (including what adventuresome or ignorant parents try out on their children) will be a rocky road—but eventually, emerging from the serious research track, there will be a science of everyday mind enhancement. Still, I'd bet on improvements in education as the major source of enhanced functioning, what makes genius more common and brings up the bottom.

Rousseau was not the first, nor even the most naïve. But he was the most famous in a line of credulous people, stretching as far back as thought and as far forward as our precarious species manages to survive, who believe that what we have left behind is better than what we have, and that the best way to solve our problems is to go backward as quickly as possible. In this view, what is past is natural, what is present, unnatural. . . .

—MELVIN KONNER, 2001

The word [Enlightenment] wasn't ill-chosen, for it bespoke illuminating a path ahead—which, in turn, implied the unprecedented notion that "forward" is a direction worth taking, instead of lamenting over a preferred past. Progress—and boy, did we take to it. In two or three centuries our levels of education, health, liberation, tolerance and confident diversity have been momentously, utterly transformed.

—DAVID BRIN, 2002

The Major Transitions in Evolution
adapted and **expanded** from those of Maynard Smith and Szathmáry, 1999

1. Bagging those replicating molecules inside a cell membrane.
2. Centralizing replicating molecules onto chromosomes.
3. The division of labor between DNA's information storage and RNA's construction activities.
4. A beyond-the-bacterium cell, the live-together-or-die-together eukaryote confederation of organelles.
5. Sex (Don't leave variation to chance mutations: shuffle those genes with every generation).
6. Making various specialized cells from the same DNA.
7. From solitary cells to coexisting in groups (about a billion years ago).
8. **From primate socialization to having protostructure and protolanguage abilities (perhaps a million years ago).**
9. **From unstructured short sentences to coherent higher intellectual functions (the transition to the modern mind was perhaps 50,000 years ago).**
10. **The superhuman transition. (Pending?)**

12

What's Sudden
about the Mind's Big Bang?

The moderns somehow got their act together

MAJOR TRANSITIONS DON'T HAVE to be fast, only profound. They are something like phase transitions (ice to water, water to steam)—revolutionary, not just evolution as usual.

Many of the past major transitions are about coming-together successes, rather like corporate mergers. Most happened more than a billion years ago. The last transition is usually stated as simply language or the "mind's big bang." I subdivide it, myself, into proto- and then the real thing; still, they are both in the last million years or so.

Protostructure and protolanguage probably developed slowly. Was the behaviorally modern transition from the proto version particularly quick?

Maybe speed doesn't matter. The archaeologists still might show that it was really spread out in stages. And speed is sometimes merely in the eye of the beholder: if your unit of time is geologically slow, then something that takes a few million years (such as the demise of the dinosaurs) looks quick. Get data with better time resolution, and you may discover a gradual ramp or a series of events.

But speed does matter, should two processes be operating in parallel or in opposition. That's because the faster one becomes what later processes build atop. Absolute speed might not matter, but *relative* speed often does. So of all the candidates for the last transition so far, which ones have the right stuff to be the quickest?

THE first thing that I was taught by my high-school journalism teacher was to always cover, in the course of writing a story, the five W's and the H: *Who, what, when, where, why,* and *how.* This brief history of the mind is a summary of all of them, but *why* and *how* are sometimes a little tricky, given our usual tendency to focus on one "cause" at a time and suppose that it is in opposition to other candidates. Or to assume that they are chained in sequence. (Of course, all might be needed, like the stones in an archway.)

Why and *how* are both about the process by which one thing turns into something else. *How* usually focuses on the here-and-now mechanics of the change, while *why* usually directs our attention to the setup phase and provides a rationale for things operating as they do now. Yet *why* questions are usually just *how* questions operating on a longer time scale, and focusing on whole populations rather than the individual.

Among anthropologists, the first *why*'s that come to mind are social. Increasing group size allows wider cooperation and occupational specializations—but it also challenges you to keep track of who-owes-what-to-whom. Gossip might become very important, a payoff for protolanguage. Animals that live in larger social groups have bigger brains.

A growing cultural toolkit, both vocabulary and staged toolmaking, can build combinations—but then you have to judge the coherence amidst more ambiguity. Just as the invention of money vastly expanded the barter economy, so researchers have suggested a similar expansion for social stuff that falls short of syntax itself. I agree they are reasonable, that such things could be like a common currency being invented. My question is whether something else got there first. (That's often the issue in evolutionary arguments. There may be a number of reasonable candidates, things that ought to have been useful, but some are surely slow and no one can yet judge which is the fastest track.)

W E now have a few *how* candidates for the mind's big bang, things that operate on the neurophysiologist's favorite time scale, the milliseconds-to-minutes span of thought and action:

A. The secondary use of throwing's structured planning facility for other planning in other modalities and on longer time scales. Maybe the transition is when a major secondary use developed.
B. More effectively managing creativity's incoherence via a Darwinian process that improves quality, so that you "don't go off half-cocked." Maybe the transition is when the quality finally improved enough so that the surviving novel combinations were more useful than dangerous.
C. The maintenance of independent planners by cortical partitioning, so that all of those phrases and clauses can maintain their separate identities and competitions while still playing a role in the overall plan/sentence.
D. Spatiotemporal firing patterns that can circulate undistorted between cortical areas ought to be a big improvement for handling novelty on the fly. Maybe the transition is when error correction gets good enough to transmit codes without the usual distortion that makes everything a special case to be learned over a number of repetitions—fine for learning but not for first-time novelty.

And on a different level of explanation, that of the childhood development time scale's *how* and the many-generations evolutionary *why*, we
have an "EvoDevo" candidate for the first appearance of whichever the
crucial one is, something capable of producing a runaway acceleration
into the creative explosion:

E. If children exposed to structured stuff can softwire their
   brains to better handle it, and if the younger they are exposed,
   the better they do as adults, then the more precocious children
   will soon double the amount of structured speech heard by
   the next generation of youngsters. Some of their children will
   be even more precocious, and so become even better as adults.
   In this way, the typical age of acquisition of structured stuff
   might plummet from eight years (tuned up by throwing) to
   three years (tuned up by spoken language) because they can
   hear (and see) novel structured examples long before their
   fine movement coordination is capable of practicing novel
   ballistic movements. So the transition might be language, acting like a contagious disease over a few generations' time.

Undoubtedly more candidates will be forthcoming on each time scale
and at other levels of organization, but these five will do to illustrate
the task of analyzing the *how* and *why* candidates for their contributions to the transition we call the mind's big bang.

"ESSENTIAL but not sufficient" probably operates here. They all
may have been essential for being behaviorally modern but only
one may have changed rapidly at the transition and finally made possible the flowering of the modern mind. The question is not when the
last one was added to the archway but which has the growth curve that
becomes steeper and steeper because things build on themselves.

A word about linear versus exponential growth. In linear growth, nothing changes the base on which the next round operates. If the truck factory's annual output grows from 50 to 100 trucks in 25 years, there is nothing that says it will redouble in the next 25 years. If the trend continues, it may just go from 100 to 150 trucks annually—not redoubling to 200, because the existing trucks do not (we hope) beget more trucks.

But suppose that the average mother doubles her output of children surviving childhood from, say, two to four. If her girls are just as successful at parenting, there are eight children after another 25 years. And another redoubling to 16, and 32, and so on. When the next generation is some multiple of the present generation, it is called exponential growth over the generations. When the average mother has three children, as in Kenya today, the population doubles in about 23 years if the death rate and emigration do not change. You see the same exponential growth in epidemic disease spread: more active cases means even more people exposed in the next round. The incidence can keep doubling and redoubling every month until acquired immunity or isolation procedures slow down the growth rate.

There are a lot of things potentially involved with the expansion of the human mind. Some grow over time. But do they grow exponentially, because the base-on-which-to-grow itself increases in the next generation? Does success breed even more success?

So now let me ask which of the aforementioned candidates has the right stuff, an exponential growth curve that might look like a fast transition rather than like ordinary improvements:

A' The secondary use of ballistic movement planning circuitry for other things looks gradual to me. More types of use, on more and more occasions, and on longer and longer time scales—but maybe not a very steep growth curve. By itself (but see E').

B' Making creativity "good enough" might indeed have a course of hidden improvements, where plans become more coherent but still aren't safe enough to act on. But after they get good enough to pass this threshold, the growth looks like more and more things, more and more of the time.

C' Partitioning the cortex dynamically ought to look incremental.

D' There are threshold aspects to the ability of one cortical area to talk to another using error-correcting codes, avoiding the necessity of learning a special case for every concept to be communicated over long distances. This one looked pretty good to me when I wrote *The Cerebral Code* in 1996. But corticocortical codes could become more efficient, one pathway at a time. You'd just get better and better as more areas got the trick of making their pair of interconnections use error-correction features and a common code. Yes, it grows more steeply because of the two- and three-hop possibilities, but I still suspect that it builds on itself only at a moderate pace.

So none of the physiological-time-scale candidates has an obvious positive feedback that makes the growth curve get steeper and steeper over time. But the EvoDevo candidate looks to have great possibilities for explosive growth, since there is presumably a reservoir of capable but unexposed children just waiting for culture to bring structured examples to them early enough in childhood.

E' The more precocious children do better as adults and leave more variants around, some of which are even more precocious than they were. And so, over dozens of generations, even precocious two-year-olds might be successfully infected with syntax. Each mother speaking a structured language rather than protolanguage serves to "infect" a number of children growing up within hearing distance. And when they them-

selves have children.... You'd see an explosive growth, both locally and via lateral spread, in structured language users on the
thousand-year time scale.

Precocious kids doing softwiring better, furthermore, serves as a good
example of how cultural developments were likely essential—why one
cannot make the usual false dichotomy between genes and culture. More
culture selects for the genetically more precocious of the next generation.

Indeed, any gene changes might merely be in what promotes precocious acquisitiveness of words and their patterns. Behavior invents,
anatomy follows via gene tweaks. But here the anatomy isn't gross size
(what we can measure in fossil skulls) but microscopic change affecting
the plasticity of synapses between neurons. Nothing in this partial list
of candidates suggests that a bigger brain might be needed for the final
coming together of the committee.

Does this list of five candidates summarize how we got structured
higher intellectual functions with on-the-fly creativity of quality? Is the
rest of this story "just applications"? Or has this mental machinery also
continued to evolve since it emerged in the middle of the most recent
ice age? The story about how reading specializations are softwired into
the brain shows how new cognitive tasks can achieve anatomical specializations during life. And my EvoDevo argument suggests how Darwin's inheritance principle can gradually change acquisitiveness.

Not everything that culture invents will help softwire the brain in
youth. But reading specializations certainly show that culture in childhood can sometimes build a more capable adult brain. It makes you
wonder what the next round of educational improvements and cognitive challenges will do in softwiring the brain.

IF language spreads like an infection, there is a useful distinction that
the epidemiologists make that might help us think about the conservatism of what came before *Homo sapiens sapiens.*

Genes are not the only thing that a mother passes on to her off-spring. The hepatitis B virus may be passed on as well. So-called vertical transmission is like inheritance in that you get it from your relatives. In horizontal transmission, you get it from unrelated persons.

Vertical transmission is slow. After all, it takes a whole generation's time to pass it on. Things passed on this way, which include a number of cultural traits such as food preparation and the non-expert aspects of toolmaking, cannot change very rapidly. The history of toolmaking, from 2.6 million years down to 150,000 years ago, is one of conservatism and only occasional innovation.

In horizontal transmission, things can be copied, mimicked in an instant's time. Because there are many copying errors, there will soon be a number of versions being practiced. Some will be better and will themselves be preferentially copied. Serious mimicry and real teaching make horizontal transmission of culture work even better.

Reciprocal altruism is a form of horizontal transmission that evolves atop those within-the-family forms of sharing. As Luca Cavalli-Sforza points out, sharing gets a big boost from language. (Even, I would add, from those short sentences of protolanguage.) It is expensive to share food—there are immediate costs to the giver—but sharing information is cheap. You might not want to share your knowledge of the best fishing hole, but telling someone about an ample resource—or relating who did what to whom in their absence—has little cost. When you can create long sentences without the protolanguage equivalent of an elaborate charade, information sharing gets a big boost.

The spread of culture can, of course, modify genes. For example, milk from grazing animals is a nice supplement to the diet but if the enzyme that helps digest milk works only up to the age of five or so, then this food source is useful only to young children. But there are variant genes that prolong this period of being able to digest milk. Those variants have become very common in northern Europe where reindeer milk was an important food source in winter. They have become common in some parts of Africa but not others, depending on the local cultural practices with regard to adults drinking milk.

The general principle, remember, is that behavior invents and adaptation via gene changes makes the invention more efficient. New form follows new function. Even if the invention of beyond-throwing structured stuff did not take new gene combinations, as in marching up that earlier-is-better curve, one would expect a series of subsequent genetic changes to make it more efficient.

Up-from-the-apes causes are numerous, and the five candidates for the 50,000-year transition that I have discussed were selected to show speed considerations.

But they are only the present foreground considerations, and there are background considerations that I have mentioned that would be another writer's foreground. I am particularly impressed with adding teaching and enhancing mimicry (see page 73) as important aspects, and the role of beyond-the-apes attention—both joint attention and the addition of the hunter's versatile attention span to the ape repertoire.

The "hundred" differences between apes and humans are all essential aspects of being human. They are not likely to line up in some chain of cause and effect. A web of push and pull is more likely, and attempts to identify the fast tracks must be viewed as only part of the explanatory attempt.

T he big thinkers in the sciences of human nature have been adamant that mental life has to be understood at several levels of analysis, not just the lowest one. The linguist Noam Chomsky, the computational neuroscientist David Marr, and the ethologist Niko Tinbergen have independently marked out a set of levels of analysis for understanding a faculty of the mind. These levels include its function (what it accomplishes in an ultimate, evolutionary sense); its real-time operation (how it works proximately, from moment to moment); how it is implemented in neural tissue; how it develops in the individual; and how it evolved in the species.

—STEVEN PINKER, 2002

13

Imagining the House of Cards

Inventing new levels of organization on the fly

L EVEL OF ANALYSIS IS an unavoidable concept in biology, where the process of coming into being is so varied. In his history of biological thought, Ernst Mayr distinguished between proximal causes (physiological stuff, such as the mechanics of brain operation) and ultimate causes (the evolutionary setup phase, what makes the proximate mechanisms what they now are). Our present distinction of development (what children do) from evolution (what species do) seems natural but, in Darwin's day, the terms were different. Indeed, Darwin didn't much employ the term "evolution" as, back then, it simply implied a pattern unfolding, as in a dance or a coordinated military maneuver (what marching bands now do at halftime). So evolutionary setup, mechanics, and development are the main levels of analysis (or explanation) in biology.

A level of organization is a more general concept, seen in all of the sciences. This kind of level is best defined by certain functional properties, not anatomy. As an example of four levels, *fleece* is organized into *yarn*, which is woven into *cloth*, which can be arranged into *clothing*. Each of these levels of organization is transiently stable, with ratchetlike mechanisms that prevent backsliding: fabrics are woven, to prevent their disorganization into so much yarn; yarn is spun, to keep it from backsliding into fleece.

Neuroscientists once talked at cross purposes when arguing about learning—is it an alteration at the level of gene expression, ion channels, synapses, neurons or circuits? All of the above? Confusion of levels also occurred in evolutionary science in the wake of Darwin when geneticists in the 1900s thought that the newfound genes were an alternative explanation to natural selection. It is common to initially suppose that complementary causes are, instead, competing explanations.

A proper organizational level is characterized by "causal decoupling" from adjacent levels; it's a "study unto itself." You can weave without understanding how to spin yarn (or make clothing). Indeed, Dmitri Mendeleev figured out the periodic table of the elements without knowing any of the underlying quantum mechanics or the overlying structural chemistry. Most of the natural sciences need only several levels of organization. There are, however, at least a dozen levels of organization within the neurosciences—all the way up from genes for ion channels to the emergent properties of cortical neural circuits. And, if we invent a metaphor, we tack on a new level. Then there are those developmental and evolutionary levels of explanation.

The closest approximation to a word, in the animal world, is an emotional utterance such as the chimpanzee's "What's that?" or "Get away from that!" equivalents. Occasionally they can be interpreted as nouns ("snake" or "eagle"). We humans can combine several utterances for an additional meaning, say, "That's big." This opens up a space of thousands of new meanings. In addition to such relationships, we can compare items, say, "This is bigger than that." We can even build a new level, that of relationships between relationships, when we say "Bigger is better." It is this on-the-fly construction of a new level, as when we find an analogy or use figurative speech, that makes human cognition so open ended, totally unlike anything seen elsewhere in evolution. Nothing in animal communication is in this class.

IT is only a Seattle coffee joke, but it nicely illustrates different levels of mental organization.

The ascent to higher levels of consciousness begins when you first contemplate the toothpaste in the morning, when you can operate only at the level of single words and well-memorized actions.

Relationships between concepts, like speaking in sentences, may first require priming, with the morning cup of coffee.

Talking about relations between relationships (better known as analogies or metaphors), as when we say "Bigger is better," may require a double espresso.

Poets, of course, invent figurative speech and compound it into blends. ("The path not taken.") Some seem to require a series of stage-setting maneuvers involving many superstitious practices, some of which involve substances even more toxic than coffee.

MENTAL life can pyramid a number of levels, thereby creating structure. We see the pyramiding of levels as babies encounter the patterns of the world around them. A baby first picks up the short sound units of speech (phonemes), then the patterns of them called words, then the patterns within strings of words we call syntax, then the patterns of minutes-long strings of sentences called narratives (whereupon she will start expecting a proper ending for her bedtime story).

By the time she encounters the opening lines of James Joyce's *Ulysses*, she will need to imagine several levels at once: "Stately, plump Buck Mulligan came from the stairhead, bearing a bowl of lather on which a mirror and a razor lay crossed. A yellow dressinggown, ungirdled, was sustained gently behind him by the mild morning air. He held the bowl aloft and intoned: Introibo ad altare Dei."

As always, there is the physical setting (piecing together the top of an old Martello gun tower overlooking Dublin Bay with a full-of-himself medical student about to shave).

But the ceremonial words and deliberate pace prompt you to consider the more abstract level of metaphor. An ungirdled gown and an offering of lather? Is this—gasp!—an obscene mockery of the Catholic mass, far more blasphemous than anything that Salman Rushdie might have implied about Islam? And Joyce is instead celebrated in Ireland?

So much of our intellectual task, not just in reading Joyce and Rushdie but in interpreting everyday conversation, is to locate appropriate levels of meaning between the concreteness of objects and the various levels of category, relationships, and metaphor. You usually cannot get the joke without locating the correct level of organization to which it refers, and it is often the alternative interpretations at different possible levels that makes it so funny.

"And about how many people work here?" the visitor politely inquires of the boss.

"About half."

O UR minds can operate on the unreal ("the missing chair" or "zero") but it is tropes that allow us ways of saying "this is like that." They tack on imagery with connotations above and beyond the literal meaning. We employ spatial metaphors such as "soaring spirits" and "falling GNP." It is claimed that much of learning is dominated by analogy (the heart is *like* a pump) and metaphor (the heart *is* a pump). Roland Barthes declared that "no sooner is a form seen than it must resemble something: humanity seems doomed to analogy."

Even in science, extended metaphors personify by giving human characteristics to charged particles where those of like charge "hate" one another though those of opposite charge "love" one another.

Metaphor is the first thing we try when working our way into a complex subject, but in doing science you eventually try to replace it with something better (though the metaphor may remain handy for teaching). In seventeenth-century England, the scientists of the Royal Society sought "to separate knowledge of nature from the colours of rhetoric, the devices of the fancy, the delightful deceit of the fables." They saw the "trick of metaphors" as distorting reality. Yes, but something is better than nothing.

To keep creative constructions from being nonsensical, two tasks are needed. The first, as mentioned earlier, is to judge new associations for their internal coherence: do they all hang together in a reasonable, safe way? (Initially, most associations are surely as incoherent as our dreams, which provide us with a nightly experience of people, places, and occasions that simply do not fit together.) Awake, it's an off-line search for coherence, for combinations that "hang together" particularly well. Sometimes this provides an emergent property: the committee can do something that none of the separate parts could manage.

Second, to spend more time at the more abstract levels in an intellectual house of cards, the prior ones usually have to be sufficiently shored up to prevent backsliding. Poets, in order to compare two candidate metaphors, have to build a lot of scaffolding. Finding the right combination can be like adding a capstone to an arch, which permits the other stones to support themselves—as a committee, they can defy gravity and dispense with the temporary scaffolding, so slowly assembled with the aid of the writer's superstitious rituals and self-medications.

The loss of a normal adult ability to locate and hold a level of organization is what the psychiatrists are testing for when they ask you the meaning of a proverb like "People who live in glass houses shouldn't throw stones." Without being able to stabilize your house of cards at an intermediate level, you can't reason by analogy. A psychotic patient may be both very concrete and hopelessly abstract, as if flipping from one extreme to the other because unable to settle at the mezzanine for any length of time.

Finding the appropriate level at which to address a problem—not too concretely, not too abstractly—is an important aspect of intelligence that is probably not seen in the great apes. Searching the wrong level is a common blunder in all of higher intellectual function—when you don't "get it," it is often because you cannot locate the intended level of reference where everything falls into place.

We can distinguish "John believes there is a Santa Claus" from "There is a Santa Claus" (if apes ever manage this, it will surprise a lot of researchers). We usually do it so well that a persistent breakdown of this ability raises the question of schizophrenia. Still, most of us easily confuse concepts that can be approached from different levels of analysis, and sometimes we get stuck rationalizing the results—as when a statement of moral principle (say, equity feminism's equal opportunity and equal pay) gets confused with demonstrable-but-irrelevant facts about biology (say, when some gender feminists get upset about reports of brain differences between males and females and worry that needed social progress will suffer unless this heresy is vehemently denied).

> One of the reasons we so seldom paint ourselves into a corner or saw off the limb we are sitting on is that we have all heard one funny, memorable tale or another about a chap who did just that. And if we follow the Golden Rule, or the Ten Commandments, we are enhancing our underlying natural instincts with prosthetic devices that tend to encourage framing the situations we confront in one way or another.
>
> —DANIEL DENNETT, *FREEDOM EVOLVES*, 2003

A sense of ethical behavior has some foundations in primate social life, where even a monkey has a sense of what he can get away with and what will cause trouble. Framing and a "theory of mind" add much more capability.

We use our theory of another's mind all the time in conversation when we pitch a sentence in a way that takes account of our listener's knowledge. I say "brain cell" if I'm not sure my listener knows the term "neuron." When you say "he" instead of John Smith, or "the" instead of "a," you are implying that what you mention is something that your listener should know already, because of shared knowledge or some antecedent communication.

This everyday practice of estimating what another knows or believes helps you acquire an ability to put yourself in someone else's shoes. I've saved the topic for here, late in my brief history of mind, because I suspect that our kind of ethics probably owes a lot to simulating novel social situations in your head before acting. And you often operate at a second or third remove: I may jaywalk when there aren't children around, but I will detour to the crosswalk when I might be serving as a role model. I may avoid saying something to X because I can estimate that X wouldn't want Y, nearby, to overhear it because Y would probably tell Z.

Of course, it cuts both ways: if your competitiveness exceeds your empathy, you may use your mental model to take advantage of the other person rather than help him avoid a problem—as when trying to outthink your chess opponent by projecting what the board might look like, three moves ahead. Pretense involves treating some things as a harmless game, when competitiveness and deception within the rules can override other considerations. But they are exceptions carved out of a broad area in which you are expected to tread lightly, to minimize the impact on others of your moves.

Some of this may have been around long enough to hardwire some intuitions into the brain. We have some emotional responses such as embarrassment, shame, and guilt that are not shared with the other great apes. (Nor are the new emotions commonly featured in our dreams in the manner of, say, anxiety.) As Mark Twain said, man is "the only animal that blushes. Or needs to." And the social setting—as in that chimp patrol—can transform otherwise peaceful, thoughtful individuals into irrational, suggestible, and emotional brutes.

The new emotions suggest, to me, the role of a good reputation in future social dealings. That they have become instincts or intuitions makes me suspect that some protostructure and protolanguage might have been present in social life for a long time, well before the last transition.

W E have achieved an extraordinary ability to pretend, fantasize, lie, deceive, contrast alternatives, and simulate. But levels are the real stuff of creativity, so let me give an appreciation of one of the greatest feats of creativity: the everyday emergence of new levels of organization.

Here is an example of two input spaces serving to prompt you to construct a third hybrid space in your mind. It is from a sailing magazine reporting on a "race" between two boats whose journeys were actually 140 years apart: "As we went to press, Rich Wilson and Bill Biewenga were barely maintaining a 4.5 day lead over the ghost of the clipper *Northern Light*, whose record run from San Francisco to Boston they're trying to beat. In 1853, the clipper made the journey in 76 days, 8 hours."

We deal easily with such metaphorical constructions, mapping the old journey onto our trajectory planning for the modern trip to create a "ghost" lagging behind. Understanding one story by mapping it onto a more familiar story (that's what constitutes a parable) shows how we can operate mentally, once we have the structure for syntax and can use it again for even more abstract, beyond-the-sentence constructions. We map actions between the spaces but perhaps substitute new actors. We do something similar in logical reasoning.

Blended spaces draw from several source frames that are closer to reality. The resulting "blend" inherits qualities from each input but often achieves some unique properties of its own. The blending process lets you suggest to your listener that there are connections between elements, even though their properties may be materially different. Blending says a lot about our creativity, as in this description by Mark Turner in *The Literary Mind*:

> Certainly there is considerable evidence that blending is a mainstay of early childhood thought. A two-year-old child who is leading a balloon around on a string may say, pointing to the balloon, "This is my imagination dog." When asked how tall it is, she says, "This high," holding her hand slightly higher than the top of the balloon. "These," she says, pointing at two spots just above the balloon, "are its ears." This is a complicated blend of attributes shared by a dog on a leash and a balloon on a string. It is dynamic, temporary, constructed for local purposes, formed on the basis of image schemas, and extraordinarily impressive. It is also just what two-year-old children do all day long. True, we relegate it to the realm of fantasy because it is an impossible blended space, but such spaces seem to be indispensable to thought generally and to be sites of the construction of meanings that bear on what we take to be reality.

Levels and blending do make you realize that understanding the underlying neural processes has enormous potential for further enhancing our cognitive processes—that the last transition might someday become only the penultimate transition.

Only now are we beginning to sense a hinge in history, a time when the earth is beginning to move beneath our feet. In the near term [of an exponential increase in technology affecting human capabilities], the world could divide up into three kinds of humans. One would be the Enhanced, who embrace these opportunities. A second would be the Naturals, who have the technology available but who, like today's vegetarians, choose not to indulge for moral or esthetic reasons. Finally, there would be The Rest—those without access to these technologies for financial or geographic reasons, lagging behind, envying or despising those with ever-increasing choices. Especially if the Enhanced can easily be recognized because of the way they look, or what they can do, this is a recipe for conflict that would make racial or religious differences quaintly obsolete.

—JOEL GARREAU, 2003

There is no more powerful law of nature than that of unintended consequences. However carefully we might think out the possible results of our actions, they are likely to give rise to difficulties we hadn't thought of—and fixing secondary problems of our own making is often more difficult than addressing those presented to us by Nature.

—IAN TATTERSALL, 2002

14

The Future of the Augmented Mind

A combustible mixture of ignorance and power?

ARE THERE GENETICALLY ENGINEERED prospects of super genius —maybe even a do-it-ourselves successor species to *Homo sapiens sapiens*? Or some lash-up of computers and people that will create a hybrid?

Not soon, I suspect—and to get to the long term, civilization has to first survive the short term. (For example, how would we avoid genocides in the transition period?) I have already expressed, in chapter nine, my doubts about the course-plotting skills of the "not ready for prime time" prototype that escaped prematurely from the African cradle and took over the world.

Most popular speculations about mind's future (say, those mind-liberated-from-body enthusiasms featured in slick magazines advertising cooler-than-cool gadgets) lack any cognitive perspective on the limitations of our prototype. Nor do such articles seem to offer any anthropological perspective of the evolutionary trajectory we've been on. Nor any neurobiological concern with the stability problems already evident in seizures and mental illness. The implications of the growing divide between The Enhanced and The Rest is seldom addressed.

This brief history of the mind is not the place to critique these blinkered views of the future, nor the place to sketch out why genetic manipulations may not turn out to be quite what we would hope. By offering a rather low-tech glimpse of the future, I can focus on patching up the prototype and addressing what we will build atop it. This final chapter is not a speculation about specific futures, though I will mention some cautions.

Scientists have no special wisdom in areas of ethics and stewardship, just a strong tradition of skepticism and theorizing. And, in common with the technologists such as Bill Joy, we sometimes have the knowledge from which to give early warnings of trouble ahead. That's different from knowing what to do about it, or where wisdom lies.

But we do have a major responsibility, to get across to more general audiences and policymakers the nature of the challenges coming up, so that initiatives can properly focus on the long term. Some things really are important and it has proven easy to lose sight of them in the gee-whiz version of the future.

Unfortunately, no one seems able to discuss the future of the mind without marveling about this exhilarating speed of technology and the power of human-computer hybrids. They do tend to grab the attention. So perhaps I should first offer some perspective about the setting in which mind's future might unfold—graying, speedups, wireheads, pumping up IQ, and emergent properties more generally—before I tackle the properties of future mind per se.

ONE of the more pervasive changes in average mind may occur because the average mind becomes much older. More experienced and less prone to beginners' mistakes, perhaps, but likely far less energetic and adventuresome.

I was recently asked to imagine a day in my life, assuming that I lived until the year 3000. I thought about walking around the neighborhood on my artificial hips and my artificial knees. I suspected that my mood would be sad. I would be thinking about how we had dug ourselves into a deep hole, and that it would be hard to escape from it.

On average, people turn more and more conservative with age, self-centered and disinclined to rock the boat. By the year 3000, I would be experiencing the loneliness of the last liberal.

EXTRAPOLATING speed is easy to do and exponential growth curves abound, such as the Moore's law doubling of memory chip capacity and processor speeds every several years over the last few decades.

Many technology aficionados suggest that as exponential technological change continues to accelerate into the first half of the twenty-first century, "it will appear to explode into infinity, at least from the limited and linear perspective of contemporary humans," as the inventor Raymond Kurzweil says of "the singularity," resulting in "technological change so rapid and profound that it represents a rupture in the fabric of human history." Like Malthus and the population bomb, it's possible—unless something else slows it down before it rips the social fabric. Or uses the new technological capabilities, in the manner of Aum Shinrikyo in the sarin attack in Tokyo, to hasten some rapture-promoting Armageddon.

Things usually happen in the meantime to interrupt or rechannel exponential growth. Recall those 1950s extrapolations of leisure time, where the wage earner would get a shorter and shorter work week. Back then, one salary often supported an average family of five. And what happened to this vision of enhanced leisure time? Now it takes two salaries to support a family of four. Someone forgot about the Red Queen principle in Lewis Carroll's *Through the Looking Glass*:

"Well, in *our* country," said Alice, still panting a little, "you'd generally get to somewhere else—if you ran very fast for a long time, as we've been doing."

"A slow sort of country!" said the [Red] Queen. "Now, *here*, you see, it takes all the running *you* can do, to keep in the same place. If you want to get somewhere else, you must run at least twice as fast as that!"

Our technological lifestyle has already begun changing so rapidly that a person's working lifetime has to include one career after another after another. But only the best and the brightest can cope with such frequent retraining, leaving most of the population constantly battered by insecurity and lack of job satisfaction, alienated by the situation in which they are trapped.

So there are big problems with the speed of change. The faster you go, the more easily a pothole can spin you out of control. But as I earlier noted, it usually isn't speed by itself that matters; it is *relative* speed. Army generals love the blitzkrieg concept, of overrunning the enemy before it can effectively react. People (and societies) can overrun themselves, too. Your own speed of travel must be judged relative to your speed of reaction. If you can't shorten your reaction time commensurate with your faster speed, or cannot find better headlights to give you a longer view, then things that would give you no trouble at normal speeds can give you a lot of trouble at higher speeds.

And reaction times are only the simplest application of speed differences. When innovation operates in one area faster than in related ones, when one is nimble and the other is ponderous, things can bend and break. Contrast the speed of technological advance with that of societal consensus. It took less than a decade to put together an atomic bomb, once the physics was understood. It took only four years after the first free web browser appeared until there were a billion web pages worldwide, indexed by a free search engine that even children could use without formal instruction. Compare those technological spurts to one of the best examples of political progress, short of shaky revolutions: the European Union took 50 years, two generations of politicians, to get to the stage where the Euro started to circulate. And that's fast for consensus building.

The science and technology of mind may move far more quickly than we can create consensus about what to do—say, for insuring that things go on, that individuals' independence and upwards mobility in society is maintained, that costs and benefits are distributed, that stratification does not develop in society and become a caste system.

We do get better headlights from science—something increasingly important as things speed up—but the political reaction times are so slow that it hasn't helped much. It has been clear for at least 30 years that greenhouse problems were upon us, but denial still reigns in high places (so too does ignorance of science). The same nimble-ponderous problem will likely be seen as the future of the mind unfolds and its societal implications become manifest.

WIREHEADS are technology enthusiasts who want to plug their brains into a computer. I'm not one of them but three decades ago, I was among the neurophysiologists who regularly wiretapped individual nerve cells in the brains of awake patients and used computers to analyze the meaning of their conversations among themselves. (It's in *Conversations with Neil's Brain*.) The technology of doing that hasn't improved very much since then. Each time some press release offers yet another photograph of a brain slice in a dish with wires attached, I get phone calls from reporters wanting to know about this exciting new prospect.

What I tell them is that I have been seeing such "news" every few years since 1964 and that, while it is nearly always competent state-of-the-art research, it hasn't yet provided much of a foundation from which to wire a wirehead—that the problems of a permanent interface are considerable, that bandwidth is still narrow (about at the Morse Code stage), and that we still don't know how to "talk" the language of the brain well enough to get across conceptual-level stuff.

The problems of doing something useful for an awake human from a carry-it-around computer are severe in the cognitive realm, though more approachable in the assisted movement applications. Someday we'll see a cognitive adjunct (probably in the area of supplementing memory, as in the "Brain in a Biceps" described in my *The River That Flows Uphill* where all that silicon memory does double-duty as a silicon augmentation of a breast or biceps). But, until we solve the interface problems, I'd bet on educational technologies. I also think improved education in the early years is what will influence far more people than either genetic engineering or the wirehead approaches.

"ALL the children are above average" in Lake Wobegon. While intended as humor, you'd think that we were heading for such an impossible utopia. While everyone tends to talk about the average as a stand-in for the whole bell-shaped distribution, the average need not change to get important effects—some of which we will surely see long before anything shifts the whole curve to the right.

Indeed, the bell curve of IQ probably began to spread early in the last century when coeducational colleges began to supplement separate men's colleges and women's colleges. So in the typical years for finding partners and settling down, the handy choices were those who could also pass the entrance exam for the college. Where you would have had more of a genetic mixing under normal circumstances between average and high (as, say, in Israel where everyone spends several years in the army before going on to college, and many mates are found in the wider choices available in this less selected population), the high-entry-requirement college tends to produce more high-high matings.

It isn't necessarily changing the average of the population. When the cream rises to the top, what's left behind is thinner—but there's no change in the milk bottle's average of fat. There are things like that, where there is no attempt at manipulating average intelligence, which nonetheless affect its distribution.

The Luke Effect (the biblical "the rich get richer") is likely to occur when parents try to assure the best for their baby via germline gene technology and by elective abortions of low-IQ fetuses. But the same exaggeration of differences can happen with education via private schools—even if the public schools catch up a decade later, you will still have an ongoing disparity, The Enhanced always well ahead of The Rest. So for at least two reasons, the IQ average just doesn't tell you what you need to know.

There's a third big reason: variability is the real stuff of evolution. There isn't a standard type, but always a highly variable population of unique individuals. The distribution is capable of being biased this way and that. But it isn't easy to engage in what we call "population thinking." It takes years to train biologists to think in terms of a variable population of unique individuals instead of a type (Platonic essences is what we default to). Without achieving that viewpoint, it can be difficult to appreciate how evolution occurs over time. "He who does not understand the uniqueness of individuals is unable to understand the working of natural selection," said Ernst Mayr.

THERE are many surprises that emerge from the intrinsically unpredictable aspects of the world. Small changes can produce big effects, and the future of mind will surely include some novelties arising from self-organization tendencies.

Some examples from the simpler world of geology: When there is a high throughput of energy, things like convection cells form. Whenever you see cliffs of basalt with hexagonal columns, remember that there are emergent properties lurking in anything that produces a steep gradient. Hot to cool may be what causes the hexagons to form (you can see it in cooking oatmeal, when you forget to stir the pan), but I can imagine softwiring emergents in the brain intensively engaging in structured stuff at earlier ages. The steeper gradients between rich and poor may produce surprising social effects unless we do something about the rich getting richer. Emergents are hard to predict, and they are not all beneficial—such as gridlock.

But many of the surprises aren't even emergents. Mentally we can invent a scheme that makes a difficult task easy. Consider trying to move a big, heavy object like the box containing a new refrigerator. You cannot lift it. You cannot easily push it across the floor because the friction is considerable. It seems an impossible task. But in trying to maneuver it, you discover the technique of walking it across the room: you tilt it back onto the near edge, then rotate it around one corner, then the other corner, "walking" it across the room with little more effort than it takes to keep it on edge. It is much like sailing into the wind at an angle, tacking back and forth—something else that initially seems counterintuitive. Our mental life often makes such shortcut discoveries on more abstract levels, and we might get even better at it in the future.

So much for the general principles and the gee-whiz settings that usually distract us. What about the properties of future mind per se?

Speculation is never a waste of time. It clears away the dead-wood in the thickets of deduction.

—THE NOVELIST ELIZABETH PETERS, 2000

W HERE does mind go from here, its powers extended by science-enhanced education and new tools—but with its slowly evolving gut instincts still firmly anchored to the ice ages? With the mental hardware still full of the shortcomings of the rough-around-the-edges prototype, the preliminary version that evolution never got a chance to further improve before the worldwide distribution occurred?

Perhaps we will come to manage our minds better, as some Buddhists aspire to do, learning how to put things on the back burner and revisit them, rather than worrying continuously.

Evolutionary psychiatry will probably give us some alternative ways of looking at common disorders—and perhaps offer us some paths to improving mental functioning. Mood disorders like depression are, of course, the most common of problems, exerting a pervasive bias on what interests us. Of all the mental illnesses, depression is the easiest to appreciate as an evolutionary adaptation as it seems widespread in mammals. A wounded animal holes up, doesn't move much, loses its appetite and interest in sex—all of which makes perfect sense if there is a broken bone or wound to heal.

While the mood disorders do not seem related to higher intellectual function in basic mechanism, the behaviorally modern transition and its imperfections may have made mood disorders more common in settings not part of the usual evolutionary rationale for depression, triggering the reclusive reaction when there is nothing broken. Stress is also a possible setup for depression. Adding a layer of intellect modifies this simple picture; some think that many clinical cases of depression involve the pending failure of something that the patient is emotionally committed to, that depression serves to help disengage.

Hallucinations, delusions and dementia are also the stuff of our nighttime dreams, where we see cognitive processes freewheeling without much quality control. Fortunately our movement command centers are inhibited during most dreams, so we don't get into trouble acting on dangerous nonsense. When similar incoherence is the best thing our consciousness has available during waking hours, it may be part of a thought disorder.

Obsessions and compulsions are lower-level stuff, somewhere between thought and mood disorders, but they seem related to agendas and their updating. Our cat may have some instincts for keeping track of the field mice in the back yard and revisiting each of our closets every few weeks, but behaviorally modern humans have very versatile, structured agendas. Yet we sometimes get stuck and fail to move on, with respect to both thought (obsessions) and action (compulsions).

Just imagine the "Give him…" advertising agency able to craft an ad that causes a more normal person to obsess over the product and then go out and compulsively buy it.

R EAL mental illness may be prominent in the future of the mind. Much as I think that we will learn how to treat mental illness better, one must also consider that the number of cases might rise at the same time, perhaps just because of the speed and complexity of everyday life. And new types of malfunction may appear.

There have been several disturbing trends of late. The age of onset of major disorders has been dropping, so that psychoses are seen earlier and earlier in life. And the number of cases of autism has greatly increased in the last two decades. It takes a long time to sort out the causes of such things and, to some extent, we must suspend judgment. But the possibility of society having to cope with much more mental illness is real.

Our society has also changed in ways to make us much more vulnerable to even rare acts of mental illness. As Bill Joy said of the Unabomber, "We're lucky Kaczynski was a mathematician, not a molecular biologist." Most of the mentally ill are harmless. Those who aren't are usually too dysfunctional to do organized harm.

But I'd point out that there is a class of patients with what is called "delusional disorder." They differ greatly from schizophrenics and untreated manic-depressives because they can remain employed and pretty functional for decades, despite their jealous-grandiose-paranoid-somatic delusions. Like the sociopaths, they usually don't seek medical attention, making their numbers hard to estimate. Even if they are only 1 percent in the population (and I've seen much higher estimates), that's a lot of mostly untreated delusional people. You don't have to be mentally ill to do malicious things, and few of the mentally ill perform them, but 1 percent of sociopaths or delusional types in an anonymous big city is sure different from 1 percent in a small town where everyone knows one another and can keep tabs on the situation. And bare fists are quite different from the same person equipped with technology.

As we've seen several times in recent years, it doesn't take special skills or intelligence to create the fuel-oil-and-fertilizer bombs. Many fewer will have the intelligence or education intentionally to create sustained or widespread harm using high-tech means. But even if that is only 1 percent of the 1 percent, it's still a pool of 3,400 high-performing sociopathic or delusional techies just in California alone—and you can scale that up to the nation and world. That bad things happen so infrequently from the few Unabomber types among them isn't too comforting when the capability of that tiny fraction is growing enormously. Small relative numbers still add up to enough absolute numbers to be worrisome. With cults, you may get some warning. But here we are talking about the escalating power of the often suicidal one-person cult where deterrence doesn't work.

Fatalism, which is essentially what Bill Joy describes among the technologists, is one way of dealing with the future. But with it may go an abdication of responsibility for seeing that things go on and that everything turns out well.

It is important to distinguish between science and technology here, because the connection between them is so often oversold and simplified. Even without more new science, technology would continue producing many new ways in which society could get in trouble from unintended consequences. (The explosion of the world wide web didn't require any new science.) Science, whatever it may also do in occasionally seeding new technology, tends to provide society's long-range headlights. It is science that can detect instabilities before they cause collapse. And in combination with such technological marvels as massively parallel computers, science can provide the working models that show us the probable consequences of our actions, an important ingredient of ethical choices.

The future is arriving more quickly than it used to, and, since our reaction time is slowed by the necessary consensus building, it makes foresight more important than ever.

WILL we also shift mental gears again, into more-and-faster—juggling more concepts simultaneously, making decisions even faster? As a mundane example, consider how we struggle with remembering even 7-digit-long telephone numbers—then imagine your grandchildren able to recall 15-digit telephone numbers a day later, and even say them backwards.

It probably doesn't take genetic engineering to do this. Better training in childhood, based on understanding brains and childhood development better—as in my softwiring examples for syntax and reading—could do a great deal in preparing us to deal with more things at the same time, to hold more agendas and revisit them, and to make decisions more reliably.

Very little education or training is currently based on scientific knowledge of brain mechanisms. But that will change in the next several decades. To imagine what a difference it could make, consider the history of medicine.

Two centuries ago, medicine was largely empirical; vaccination for smallpox was invented in 1796, and the circulation of the blood was known, but scientific contributions were a tiny proportion of medicine. Digitalis was used for congestive heart failure because someone tried foxglove extracts and they worked.

Physicians often overgeneralized and it took forever to get rid of bleeding and purging. Generations of physicians were convinced that bleeding worked, but now we know it just weakened patients more quickly than the disease would have done—unless you were one of the few patients who had an iron overload disease, where bleeding could be lifesaving. Purging works for acute poisoning but not much else. A "grain of truth" is often massively misleading.

Even when they guessed correctly and avoided overgeneralization, these early physicians didn't know *how* their treatment worked, the physiological mechanism of the drug action or vaccination. When you do understand mechanism, you can make all sorts of improvements and guess far better schemes of intervention. That's what adding science gets you.

One century ago, medicine was still largely empirical and only maybe a tenth had been modified by science. It wasn't until 1896, for example, that Emil Kraepelin proposed the separation of the psychoses into schizophrenia and manic-depressive types.

These days, medicine is perhaps half empirical and half scientific (where you know not only *what* works, but a lot about *how* and *why* it works). It is only a slight exaggeration to say that the transition from an empirical to a semi-scientific medicine has doubled lifespan and reduced suffering by half.

Now consider education. Today, it is largely empirical and only slightly scientific, much as medicine before 1800. We know some empirical truths about education but we don't know *how* the successful ones are implemented in the brain, and thus we don't know rational ways of improving on them.

Yet once education has the techniques and technology to incorporate what is being learned about brain plasticity and inborn individual differences, we are likely to produce many more adults of unusual abilities, able to juggle twice as many concepts at once, able to follow a longer chain of reasoning, able to shore up the lower floors of their mental house of cards to allow fragile new levels to be tried out, meta-metaphors and beyond—the survival of the stable but on a higher level yet again.

We may expose students to the common beginners' mistakes in computer simulations, for example, so that they will become sensitized to the common logical fallacies and hone their critical thinking skills. We already do advanced versions of this; medical students now learn the consequences of not thinking ahead in simulated emergency-room situations. ("Because you didn't order a CT scan an hour ago to check for a bleeder in the brain, the patient's hidden hemorrhage has now progressed to the point of irreversible brain damage. You missed the window of opportunity to save the patient. THE END." At least it's not a real patient.) Such simulation of common errors will trickle down to educating ten-year-olds about how advertising manipulates them; done in small groups, where repeatedly getting fooled causes some embarrassment, critical thinking skills might become a more regular feature of the teen-aged mind set.

Such education, perhaps more than any of the imagined genetic changes, could make for a very different adult population. We would still look the same coming out of the womb, would still have the same genetics, but adults could be substantially different. A lot of the elements of human intelligence are things that, while they also have a genetic basis, are malleable; we ought to be able to educate for superior performance.

I think that as we move into a new generation of creative teachers augmented by teaching machines to handle the more rote aspects, they will tune into the individual's weak points and strong points. We will have children coming out of the school system who will perform very differently from the ones today—maybe not uniformly, but the high end may be substantially higher. We might bring up the bottom by more timely interventions.

Maybe those improvements in mental juggling ability will help many people think more productively, so as to head off trouble before it happens. Ethics might be a beneficiary of such improved foresight, and so might stewardship. The amount of time we spend considering the possibilities versus rushing to judgment is an example of a variable where you finally move away from making beginner's mistakes to having a much more nuanced view of things. We may be able to train for that. Our higher education pushes people in that direction, and science trains for skepticism, but there is quite a lot that we can do in childhood.

Will only the rich get smarter, or will everyone's children gain from the new flowering of education?

W HAT will happen to consciousness? And to those related things called conscience and self-consciousness? There are many subconscious aspects of mind operating in the background, such as our agendas, but in the foreground is something much more personal, the narrator of the life story capable of aspirations and reflections, capable of great achievement and pathetic meanness. To some degree, we can invent—and daily reinvent—ourselves.

And what about higher consciousness, you may ask? I'm not sure what it is (you may have noticed that I tend to talk instead about higher intellectual functions and the decision-making process), but can we jack "it" up even higher?

A great deal of our consciousness involves guessing well, as we try to make a coherent story out of fragments. The neurologist Adam Zeman lumps it all into the search for meaning: "Eye and brain run ahead of the evidence, making the most of inadequate information—and, unusually, get the answer wrong. . . . What we see resonates in the memory of what we have seen; new experience always percolates through old, leaving a hint of its flavor as it passes. We live, in this sense, in a 'remembered present.'"

The neurologist Antonio Damasio speaks of an extended consciousness having an enhanced level of detail and time span. But note that this likely could not be achieved without an equivalent in thought of syntax's past and future tenses and the long sentences made unambiguous by structuring. In short, Damasio's extended consciousness needs syntax's structuring aspect, even without overt planning or speech, just to keep mental life from blending everything like a summer drink. And to keep from getting muddled when more than maybe three concepts have to be juggled at the same time. Nor can you speculate about the future without an ability to improve novel thoughts into something of quality.

To come back to what I said at the beginning of this brief history of mind, we tend to see ourselves situated as the narrator of a life story, always at a crossroads between past and future, swimming in speculation. I think that some people today have a lot of this sense of being a narrator-in-charge, while others have less of the creative imagination needed to analyze the past and speculate about the future. In the future, we might see enhanced conscience, with the higher-order emotions like embarrassment, envy, pride, guilt, shame, and humiliation changing as well.

But at the high end, what might pump us up even higher? If our consciousness is a house of cards, perhaps there are techniques, equivalent to bending the cards, that will allow us to spend more time at the more abstract levels. Can we shore up our mental edifices to build much taller "buildings" or discover the right mental "steel?"

W E could certainly use some help, as we have some giant problems to solve soon, problems of vulnerability that civilization faces from its success. Even if we manage to fix all the rough spots and augment the higher-order stuff, we will still need to cope with two major products of higher intellectual function so far. One is population size, associated with the metaphor "the bigger they are, the harder they fall." The second is *relative* cultural speed, as in my earlier discussion about "speed kills" and "we need better headlights."

Thanks to simple planning applied to farming, population size has gone up about 6,000-fold since the beginning of agriculture. This productivity is what makes big cities possible, but we usually forget the unfortunate consequences of size. For example, if a mouse falls off a cliff, it is likely to land and get up, shake itself, and scamper off into the undergrowth. An object the size of a dog that falls off a cliff is likely to break half the bones in its body. Anything the size of a horse will splatter. To apply this to civilization, recall the earth scientists who say, "Earthquakes don't kill people, but buildings do."

Lurches can come from many things that last longer than hurricanes and earthquakes, which are over in a day and localized enough so the rest of the country can bail you out. But droughts can last for decades—far longer than the Dust Bowl of the 1930s in the United States—and affect wide areas. Some even last for centuries (North Dakota had one that lasted 700 years).

Will the farmers still be able to support 70 times their own population if we have a widespread drought? What about droughts that are both century-long and widespread? Alas, five of the last 20 centuries in North America have featured widespread droughts in the Great Plains and West that each lasted for more than a hundred years. So, just from the paleoclimate records, the present century has at least a 25 percent chance of suffering from drought conditions in which agriculture could no longer feed our large cities.

Or what happens when an agricultural monoculture gets in trouble from a widespread disease, as happened in the 1848 potato famine in Ireland? (Eliminating the seed varieties via the efficiency of a centrally manufactured seed—already a problem, which genetically modified seed will make worse—may put too many eggs in one basket.) Will the city populations quietly starve in place, or flee to further disrupt the agricultural areas?

What if such a lurch were so widespread and long-lasting that it affected much of civilization? (Worldwide droughts—usually known under the name of "abrupt cooling episodes"—have occurred many times, the average interval being 3,000 years but with the most recent one 12,000 years ago.) A collapse of civilization would not merely reduce world population size to what it was a few centuries ago. The attendant genocides during downsizings might also take us into an everyone-hates-their-neighbors hole from which it would be difficult to escape. The harder we fall, the deeper the hole we will create.

A collapse can be augmented by speed: stampedes can kill many more people than their direct cause could have done (urban panic was what Aum Shinrikyo was trying, but failed, to stir up with their Tokyo subway attacks). Our speed of communication helps set us up for panics, where a lot of people head for the door at exactly the same time.

The economic area is probably just as vulnerable to abrupt impacts as climate, and the 1997 currency crisis in Indonesia caused a lot of starvation even though food production was still working. We can now have widespread panics in the world's economies, accelerated by having 24 / 7 markets.

Our transportation systems are now moving a lot of insects and viruses around the world. Sometimes they find a new niche and are off and running. It takes time to detect them and even longer to devise effective strategies to contain the problem; this slowness of response allows a major epidemic to establish itself.

The number of people an epidemic directly kills may be only part of the problem. If you cannot get truck drivers to go into a contaminated city, a lot of people die from starvation. Lawlessness springs up and amplifies the problem. (Recall how even Baghdad hospitals were inexplicably looted when the police disappeared for a few days in 2003.)

Gradual change (as in our notions of gradual greenhouse warming) seems to be the default setting for our minds, even though evidence abounds for whiplashes. Some people assume that a free marketplace of ideas and products will solve any problem, given enough time, without realizing that many natural causes are more like a 1940 blitzkrieg invader than like a 1916 back-and-forth battlefront. We are very vulnerable to a lurch, whether from climate, disease, or economic panic. Yet we continue to treat these problems as if simple extrapolation from present-day conditions will suffice. Sustainability must also encompass surviving the lurches.

For a lurch, only a lot of organized prevention will head off the consequences—and defense is expensive, having to cover so many routes to collapse, all at the same time. (The generals say that offense is much easier than defense because you get to choose the time and route.) Judging from the past, creeping climate could suddenly turn into a blitzkrieg against civilization. Some things are too important to be left to on-the-fly improvisation and competition—and that now must include the abrupt aspects of public health, economic stability, and climate change.

THOUGH I'm generally an optimist, it is easy for me to sound pessimistic when forced to list the hazards. It's a fundamental asymmetry; a pessimist can be much more concrete about the downside than an optimist can ever be about the upside. In comparison, the possibilities imagined by an optimist will always seem fuzzy when contrasted with the known dangers having a substantial track record.

Yes, in the face of the "not ready for prime time" aspects of our intellects that I earlier mentioned, we have some serious problems. Yet much the same could have been said in earlier periods—and civilization nonetheless improved greatly in both technological and humanistic terms. As David Brin observed, "In two or three centuries our levels of education, health, liberation, tolerance and confident diversity have been momentously, utterly transformed." We cannot neglect the creeping trends and incipient lurches that endanger us, but we can also feel hopeful, given our frequent ability to transcend our apparent limitations, once we have a clear view of the challenges. Fatalism is a cheap copout.

We need to shore up civilization's foundations to deal with any type of lurch, whether climatic or economic or epidemic. And humanity has done it before: there is a famous example of shoring up your foundations called the flying buttress, and it is emblematic of our situation today.

Consider that prime example of the large-scale projects that Western civilizations have undertaken in the past: just reflect on the amount of energy and labor—the percentage of the GNP, if you like—that went into building cathedrals. And then what it took a century later, when retrofitting them with flying buttresses. This example from a thousand years ago gives us some perspective on the situation we face today, where we cannot even find the money to pay for high quality public schools and long-term projects like coping with climate change, because we are so overcommitted to less essential things.

Like some other commentators on the future, I think that we are heading into a dangerous period—full of opportunity, but precarious. We may not be gods, but it is *as if* we were—in our impact on the world and our own evolution—so perhaps, as Stewart Brand once said, we had better get good at the god business.

It's not that we need to create a new successor species, a *Homo sapiens sapiens sapiens*. But we must become far more competent at managing our situation, and become more conscientious about our long-term responsibilities to keep things going. Certainly, it is juvenile of us to think that someone else is going to clean up after us, or pick us up after we fall.

Afterword

THE DEDICATION (back at page xi) expresses my gratitude to a number of the scientists and nonscientists who have informally shaped my mental image of who I am writing for.

I have profited from many scientific discussions over the last several years as I thought through this up-from-the-apes book and would like to thank Leslie Aiello, Francisco Ayala, Paul Bahn, Liz Bates, Ursula Bellugi, Robert Bergman, Derek Bickerton, Sue Blackmore, Tom Bouchard, Luca Cavilli-Sforza, Ron Clarke, Meg Conkey, Tony Damasio, Roy D'Andrade, Iain Davidson, Richard Dawkins, Terry Deacon, Dan Dennett, Derek Denton, Ingrith Deyrup-Olsen, Robin Dunbar, Brian Fagan, Dean Falk, Tec Fitch, Pascal Gagneux, Clive Gamble, Susan Goldin-Meadow, the late Stephen Jay Gould, Katherine Graubard, Stevan Harnad, Bill Hopkins, Sarah Hrdy, Arthur Jensen, Don Johanson, Judy Kegl, Ken Kidd, Richard Klein, Mel Konner, Pat Kuhl, Kathleen Kuman, Louise Leakey, Meave Leakey, Philip Lieberman, John Loeser, Elizabeth Loftus, Linda Marchant, Alex Marshack, Sally McBrearty, Bill McGrew, Marvin Minsky, Jim Moore, George Ojemann, Maynard Olson, Gordon Orians, Steve Pinker, Todd Preuss, Robert Proctor, Sonia Ragir, Rama Ramachandran, Pete Richerson, Susan Rifkin, Giaccomo Rizzolatti, Peter Rockas, Duane Rumbaugh, Sue Savage-Rumbaugh, Oliver Sacks, Michael Shermer, Margaret

Schoeninger, Ann Senghas, Pat Shipman, Richard Stuart, Greg Stock, Mark Sullivan, Ian Tattersall, Phillip Tobias, Michael Tomasello, Gary Tucker, Ajit Varki, Johan Verhulst, Frans de Waal, Alan Walker, Spencer Wells, Nancy Wilkie, Chris Wills, Edward O. Wilson, David Sloan Wilson, Milford Wolpoff, Bernard Wood, and Richard Wrangham. The last chapter profits from futuristic discussions with Greg Bear, Stewart Brand, David Brin, Bob Citron, Joel Garreau, Wally Gilbert, Walter Kistler, Jaron Lanier, Andy Marshall, the late Don Michael, Jay Ogilvy, George Poste, Peter Schwartz, and Sesh Velamoor.

The reader can thank (as I do) Beatrice Bruteau, Ingrith Deyrup-Olsen, the late Blanche Graubard, the late Seymour Graubard, Elizabeth Loftus, Susan Rifkin, Peter Rockas, and Nancy Wilkie for volunteering to read the manuscript and pointing out my confusions and awkward moments. I hope that not too many have crept back in. The Mathers Foundation and the Whiteley Center at Friday Harbor Laboratories have helped provide the right combinations of stimulation and seclusion.

Recommended Reading

The following section and the notes are posted at
WilliamCalvin.com/BHM/readings.htm with, in some
cases, live links that will take you to the source.

There are many good books on the **paleoanthropology** suitable for
general readers. A few of my recent favorites are:

Donald Johanson and Blake Edgar, *From Lucy to Language* (Simon
    & Schuster 1996). Has excellent photographs, mostly by David
    Brill.
Ian Tattersall, *Becoming Human* (Harcourt Brace 1998).
Ian Tattersall, *The Monkey in the Mirror: Essays on the Science of
    What Makes Us Human* (Harcourt Brace 2001).
Alan Walker and Pat Shipman, *The Wisdom of the Bones* (Knopf
    1996).

Books more specifically concerned with **language** and the **transition
to the modern mind**:

William H. Calvin, Derek Bickerton, *Lingua ex Machina: Recon-
    ciling Darwin and Chomsky with the Human Brain* (MIT Press
    2000).

Terrence W. Deacon, *The Symbolic Species: The Co-Evolution of Language and the Brain* (Norton 1997).

Merlin Donald, *A Mind So Rare: The Evolution of Human Consciousness* (Norton 2001).

Robin Dunbar, *Grooming, Gossip, and the Evolution of Language* (Harvard University Press 1996).

Nicholas Humphrey, *A History of the Mind* (Simon & Schuster 1992).

Ray Jackendoff, *Patterns in the Mind: Language and Human Nature* (Harpercollins 1994).

Richard G. Klein and Blake Edgar, *The Dawn of Human Culture* (Wiley 2002).

Melvin Konner, *The Tangled Wing: Biological Constraints on the Human Spirit* (W. H. Freeman 2001).

Steven Mithen, *The Prehistory of the Mind: The Cognitive Origins of Art, Religion, and Science* (Thames and Hudson 1996).

John E. Pfeiffer, *The Creative Explosion* (Harper and Row 1982).

To read more about the behavioral commonalities seen in the **great apes**, start with:

Christophe Boesch, Hedwige Boesch-Achermann, *The Chimpanzees of the Tai Forest: Behavioural Ecology and Evolution* (Oxford University Press 2000).

Frans de Waal, *Good Natured: The Origins of Right and Wrong* (Harvard University Press 1996). Together with his other books for general readers, such as *The Ape and the Sushi Master, Bonobo, Peacemaking Among Primates,* and *Chimpanzee Politics,* you get a good view of what the ape-human transition might have been from.

Frans de Waal, editor, *Tree of Origin: What Primate Behavior Can Tell Us about Human Social Evolution* (Harvard University Press 2001).

Michael P. Ghiglieri, *East of the Mountains of the Moon: Chimpanzee Society in the African Rain Forest* (Collier 1988).

Marc D. Hauser, *Wild Minds* (Holt 2000).

Sarah Blaffer Hrdy, *Mother Nature: A History of Mothers, Infants, and Natural Selection* (Pantheon Books 1999).

Alison Jolly, *Lucy's Legacy: Sex and Intelligence in Human Evolution* (Harvard University Press 1999).

E. Sue Savage-Rumbaugh, Stuart Shanker, *Apes, Language, and the Human Mind* (Oxford University Press 1998).

Craig B. Stanford, *The Hunting Apes* (Princeton University Press 1998).

Richard Wrangham, Dale Peterson, *Demonic Males: Apes and the Origins of Human Violence* (Houghton Mifflin 1996).

For **evolution** and **cognition** more generally:

William H. Calvin, *A Brain for All Seasons: Human Evolution and Abrupt Climate Change* (University of Chicago Press 2002).

Helena Cronin, *The Ant and the Peacock* (Cambridge University Press 1992).

Richard Dawkins, *Climbing Mount Improbable* (W. W. Norton 1996).

Daniel C. Dennett, *Darwin's Dangerous Idea* (Simon & Schuster 1995).

Daniel C. Dennett, *Freedom Evolves* (Viking 2003).

Gilles Fauconnier, Mark Turner, *The Way We Think: Conceptual Blending and The Mind's Hidden Complexities* (Basic Books 2002).

Mark Johnson, *Moral Imagination: Implications of Cognitive Science for Ethics* (University of Chicago Press 1993).

Steven Pinker, *The Blank Slate: The Modern Denial of Human Nature* (Viking 2002).

Elliott Sober, David Sloan Wilson, *Unto Others: The Evolution and Psychology of Unselfish Behavior* (Harvard University Press 1998).

Michael Tomasello, *The Cultural Origins of Human Cognition* (Harvard University Press 2000).

Mark Turner, *The Literary Mind* (Oxford University Press 1996).

Edward O. Wilson, *Consilience* (Knopf 1998).

Adam Zeman, *Consciousness* (Yale University Press 2003).

For **brain side of things**, start with Deacon (1997) and my earlier books, *The Cerebral Code, How Brains Think,* and *Conversations with Neil's Brain* (with George Ojemann). Mine are all at WilliamCalvin.com.

The **glossary** in *A Brain for All Seasons* will be helpful (see William-Calvin.com/BrainForAllSeasons/glossary.htm). For language issues, see the glossary for *Lingua ex Machina* (at WilliamCalvin.com/LEMglossary.html).

Notes

When you see a short reference such as Calvin (2002),
it means that the long form is nearby or back in the
Recommended Readings.

PREFACE

XIII   R. G. Collingwood, *An Autobiography* (Oxford University Press 1939), 98.
Z. L. Henderson, "The context of some Middle Stone Age hearths at Klasies River Shelter 1B: implications for understanding human behaviour," *Southern African Field Archaeology* 1: 14–26 (1992).

XIII   There are alternative ways to define *Homo sapiens sapiens*, mostly associated with older notions of whether *H. sapiens* should include Neanderthals, about which I remain agnostic. I am simply using *Homo sapiens* for "anatomically modern" (not counting the chin) from 160,000 years ago in Africa and *Homo sapiens sapiens* for "behaviorally modern" that phases in between 100,000 and 50,000 years ago in Africa.

XIII   David Fromkin, *The Way of the World* (Knopf 1998).

XIV   Stephen W. Hawking, *A Brief History of Time* (Bantam 1989).

XVII   Daniel C. Dennett, *Freedom Evolves* (Viking 2003), 6.

1. WHEN CHIMPANZEES THINK

2   Frans de Waal, Frans Lanting, *Bonobo: The Forgotten Ape* (University of California Press 1997), 160.

3   The major references are in the list at page 194.

3   Anne E. Russon, Kim A. Bard, and Sue Taylor Parker, eds., *Reaching into Thought: The Minds of the Great Apes* (Cambridge University Press 1996).

6   Richard W. Byrne and Andrew Whiten, *Machiavellian Intelligence: Social Expertise and the Evolution of Intellect in Monkeys, Apes and Humans* (Oxford University Press 1988).

Frans de Waal, *Chimpanzee Politics*, revised edition (Johns Hopkins University Press 1998).

8   Monkeys competing with chimpanzees: Ghiglieri (1988).

8   Sonia Ragir, "Diet and food preparation: Rethinking early hominid behavior," *Evolutionary Anthropology* 9: 153–155 (2000).

8   Sonia Ragir, Martin Rosenberg, Philip Tierno, "Gut morphology and the avoidance of carrion among chimpanzees, baboons, and early hominids," *Journal of Anthropological Research* 56: 477–512 (2000) at www.unm.edu/~jar/v56n4.html#a3.

8   Mark F. Teaford and Peter S. Ungar, "Diet and the evolution of the earliest human ancestors," *Proceedings of the National Academy of Sciences* (U.S.) 97: 13506–13511 (5 December 2000).

8   Leslie C. Aiello, Peter Wheeler, "The expensive tissue hypothesis: The brain and the digestive system in human and primate evolution," *Current Anthropology* 36: 199–221 (1995).

10   Stanford (1998).

10   Richard W. Wrangham, "Out of the Pan, into the fire: How our ancestors' evolution depended on what they ate," in *Tree of Origin*, edited by Frans B. M. de Waal (Harvard University Press 2001), 121–143.

10   Craig B. Stanford, "The ape's gift: Meat-eating, meat-sharing, and human evolution," in *Tree of Origin*, edited by Frans B. M. de Waal (Harvard University Press 2001), 97–117.

10   Christophe Boesch, Michael Tomasello, "Chimpanzee and human cultures," *Current Anthropology* 39: 591–614 (December 1998).

12   Konner (2001), 40.

12   Peter D. Walsh et al., "Catastrophic ape decline in western equatorial Africa," *Nature* (6 April 2003).

2. UPRIGHT POSTURE BUT APE-SIZED BRAINS

15   The major references are in the list at page 193. For the esthetic preferences, see Gordon H. Orians, "Human behavioral ecology: 140 years without Darwin is too long," *Bulletin of the Ecological Society of America* 79(1): 15–28 (1998).

16   Michel Brunet et al., "A new hominid from the Upper Miocene of Chad, Central Africa," *Nature* 418: 145–151 (2002).

18   Peter E. Wheeler, "The foraging times of bipedal and quadrupedal hominids in open equatorial environments," *Journal of Human Evolution* 27: 511–517 (1994).

21   Hauser (2000), xviii.

3. TRIPLE STARTUPS ABOUT 2.5 MILLION YEARS AGO

22   The climate issues are covered in Calvin (2002). The beginning of the ice age at 2.5 million years is dated by N. J. Shackleton et al., "Oxygen isotope cali-

bration of the onset of ice-rafting and history of glaciation in the North Atlantic region," *Nature* 307: 620–623 (1984).

28 The D2 dopamine allele story is in Kenneth Blum, John G. Cull, Eric R. Braverman and David E. Comings, "Reward deficiency syndrome," *American Scientist* 84(2): 132ff (March–April 1996) at www.sigmaxi.org/amsci/Articles/96Articles/Blum-full.html.

31 Beatrice Bruteau, letter to author, February 2003.

4. HOMO ERECTUS ATE WELL

36 Barbara Isaac (ed.), *The Archaeology of Human Origins: Papers by Glynn Isaac* (Cambridge University Press 1989), 289–311.

36 Richard W. Wrangham, James Holland Jones, Greg Laden, David Pilbeam, and NancyLou Conklin-Brittain, "The raw and the stolen: Cooking and the ecology of human origins," *Current Anthropology* 40(5): 567–594 (December 1999).

39 Fromkin (1998), 100.

40 Carl Swisher, et al., "Latest Homo erectus of Java: Potential contemporaneity with Homo sapiens in southeast Asia," *Science* 274: 1870–1874 (1996).

40 Leo Gabunia, et al., "Earliest Pleistocene hominid cranial remains from Dmanisi, Republic of Georgia: Taxonomy, geological setting, and age," *Science* 288: 1019–1025 (12 May 2000).

40 Doreen Kimura, "Left-hemisphere control of oral and brachial movements and their relation to communication," *Philosophical Transactions of the Royal Society*, B292: 135–149 (1982).

40 Cooking: Wrangham et al. (1999).

40 The handaxe drawings by C. O. Waterhouse are adapted from Kenneth P. Oakley, *Man the Tool-maker* (University of Chicago Press 1949).

40 For the story of the Japanese monkeys, see chapter 3 in my essay book, *The Throwing Madonna* (McGraw-Hill 1983).

42 Handaxe references: Calvin (2002), 133–146.

42 Hunting per se: Matt Cartmill, *A View to a Death in the Morning* (Harvard University Press 1993).

42 Hadza hunting: James C. Woodburn, "An introduction to Hadza ecology," in *Man the Hunter*, edited by Richard B. Lee and Irven DeVore (Aldine 1968), 19–55.

42 Much of hunting in carnivores is determined by some simple innate behaviors, such as "encircle the prey" (dogs that herd animals are following this same innate tendency). The big cats clearly do not understand certain principles such as "stay downwind" and may spook their prey in a way that human hunters can avoid. See Stanley Coren, *The Intelligence of Dogs: Canine Consciousness and Capabilities* (Free Press 1994), 114–115.

5. THE SECOND BRAIN BOOM

43   The data in the figure is adapted from figure 8.3 of Richard G. Klein, *The Human Career: Human Biological and Cultural Origins*, second edition (University of Chicago Press 1999), which is based on the 1993 data collection of Aiello and Dunbar.

47   Abrupt climate changes, see Calvin (2002) and Richard B. Alley, *The Two-Mile Time Machine: Ice Cores, Abrupt Climate Change, and Our Future* (Princeton University Press 2000).

47   Law of Large Numbers: The problem with precision timing is the timing jitter. But, just as you can take four times as many independent samples to halve the standard deviation, so the brain can use ensemble averaging to reduce timing jitter. To reduce jitter eight-fold, you need about 64 times as many timers to average across. See William H. Calvin, *The Cerebral Code* (MIT Press 1996).

48   Derek Bickerton, *Language and Species* (University of Chicago Press 1990).

49   Tim D. White et al., "Pleistocene *Homo sapiens* from Middle Awash, Ethiopia," *Nature* 423: 742–747 (12 June 2003).

49   J. Desmond Clark, "Stratigraphic, chronological and behavioural contexts of Pleistocene *Homo sapiens* from Middle Awash, Ethiopia," *Nature* 423: 747–752 (12 June 2003).

50   Dunbar (1996)

51   Hauser (2000), 175.

51   Gunther Kress, Theo van Leeuwen, *Reading Images: The Grammar of Visual Design* (Routledge 1996), 168.

6. NEANDERTHALS AND OUR PRE-*SAPIENS* ANCESTORS

52   Pinker (2002).

52   Tomasello (2000), 5.

52   A readable introduction to hominid tool use is Stanley H. Ambrose, "Paleolithic technology and human evolution," *Science* 291(5509)1748–1753 (2 March 2001).

53   H. Thieme, "Lower Paleolithic hunting spears from Germany," *Nature* 385: 807–810 (1997).

54   Klein and Edgar (2002).

56   W. Tecumseh Fitch, "The evolution of speech: a comparative review," *Trends in Cognitive Science* 4: 258–267 (July 2000).

56   Philip Lieberman, *Uniquely Human: The Evolution of Speech, Thought, and Selfless Behavior* (Harvard University Press 1991). And see his *Eve Spoke* (Norton 1998).

59   Pinker (2002), 318.

59   Ian Tattersall, *The Monkey in the Mirror: Essays on the Science of What Makes Us Human* (Harcourt 2002), 168.

7. *HOMO SAPIENS* WITHOUT THE MODERN MIND

61 Max Ingman, Henrik Kaessmann, Svante Pääbo, Ulf Gyllensten, "Mitochondrial genome variation and the origin of modern humans," *Nature* 408: 708–713 (7 December 2000).

62 Michael Balter, "New light on the oldest art," *Science* news article, 283(5404): 920–922 (12 February 1999). The radiocarbon date for Chauvet is 31,000 years; the calibrated dates range from 33,000 to 38,000 years. See Edouard Bard, "Extending the calibrated radiocarbon record," *Science* 292: 2443–2444 (29 June 2001).

63 Looking modern: Daniel E. Lieberman, Brandeis M. McBratney, Gail Krovitz, "The evolution and development of cranial form in *Homo sapiens*," *Proceedings of the National Academy of Sciences* (U.S.) 99(3)1134–1139 (5 February 2002).

64 William H. Calvin, "The unitary hypothesis: A common neural circuitry for novel manipulations, language, plan-ahead, and throwing?" in *Tools, Language, and Cognition in Human Evolution*, edited by Kathleen R. Gibson and Tim Ingold (Cambridge University Press 1993) 230–250, at WilliamCalvin. com/1990s/1993Unitary.htm.

64 Rachel Caspari and S.-H. Lee, "Is old age really old? An analysis of longevity in the hominid fossil record," *Paleoanthropology Society Abstracts* (2003), at www.paleoanthro.org/abst2003.htm.

65 Leo Gabunia, et al., "Earliest Pleistocene hominid cranial remains from Dmanisi, Republic of Georgia: Taxonomy, geological setting, and age," *Science* 288: 1019–1025 (12 May 2000).

66 Blades and their African origin: Sally McBrearty and Alison S. Brooks, "The revolution that wasn't: A new interpretation of the origin of modern human behavior," *Journal of Human Evolution* 39 (5): 453–563 (November 2000).

67 Paul G. Bahn, J. Vertut, *Journey through the Ice Age*, (University of California Press, Berkeley 1997).

67 There are controversial hints of art even earlier than *Homo sapiens*: Alexander Marshack, "The Berekhat Ram figurine: a late Acheulian carving from the Middle East," *Antiquity* 71: 327–338 (1997).

67 Ofer Bar-Yosef, "The upper paleolithic revolution," *Annual Reviews of Anthropology* 31: 363–393 (2002).

67 Robert N. Proctor, "Three roots of human recency," *Current Anthropology* 44(2): 213–239 (April 2003).

67 For a discussion of the art of hunter-gatherers in the context of material culture, see Margaret W. Conkey, "Hunting for images, gathering up meanings: art for life in hunting-gathering societies," 267–291 in *Hunter-Gatherers: An interdisciplinary perspective*, ed. C. Panter-Brick et al (Cambridge University Press 2001).

67   Quote: Klein and Edgar (2002), 230.

67   Nicholas Humphrey's book, *The Inner Eye* (Faber and Faber 1986), is a good exposition on the role of social life in shaping up intelligence.

72   Michael Tomasello, *The Cultural Origins of Human Cognition* (Harvard, 2000), 32.

72   Cecilia M. Heyes, "Theory of mind in nonhuman primates," *Behavioral and Brain Sciences* 21: 101–148 (1998).

73   V. S. Ramachandran, "Mirror neurons and imitation learning as the driving force behind 'the great leap forward' in human evolution," at www.edge.org/3rd_culture/ramachandran/ramachandran_ index.html

74   Roger Shepard, *Mind Sights* (Freeman 1990).
      Ernst Gombrich, *Art and Illusion* (Princeton University Press 1969).

76   Frontal lobe functioning: I have simplified, to use an ordinary deck of cards, what neuropsychologists will immediately recognize as the Wisconsin Card Sorting Task.

77   National Institutes of Health, Office of the Director, "Early identification of hearing impairment in infants and young children," NIH Consensus Statement 11 (1 March 1993). Their recommendation is that all hearing-impaired infants be identified, and treatment initiated, before six months of age.

77   Oliver Sacks, *Seeing Voices* (University of California Press 1989), 40–44.

79   See the excellent biography by Peter Raby, *Alfred Russel Wallace, A Life* (Princeton University Press 2001).

79   Daniel C. Dennett, *Consciousness Explained* (Little Brown 1991), 21.

81   Dennett (2003), 2.

81   Daniel C. Dennett, *Kinds of Minds* (Basic Books Science Masters 1996), 147.

8. STRUCTURED THOUGHT FINALLY APPEARS

83   Richard E. Leakey, Roger Lewin, *Origins Reconsidered* (Doubleday 1992), 212.

84   Terrence Deacon, *The Symbolic Species: The Co-evolution of Language and the Brain* (Norton 1997).

85   E. Sue Savage-Rumbaugh, Roger Lewin, *Kanzi : The Ape at the Brink of the Human Mind* (Wiley 1994).

85   Sue Savage-Rumbaugh, Stuart G. Shanker, and Talbot J. Taylor, *Apes, Language, and the Human Mind.* (Oxford University Press 1998).

86   Marc D. Hauser, Noam Chomsky, W. Tecumseh Fitch, "The faculty of language: what is it, who has it, and how did it evolve?" *Science* 298(5598): 1569–1579 (22 November 2002).

86   Sentence structure: William H. Calvin, Derek Bickerton, *Lingua ex Machina: Reconciling Darwin and Chomsky with the Human Brain* (MIT Press, 2000).

92   Turner (1996).

93 Harmony: Steven R. Holtzman, *Digital Mantras: The Languages of Abstract and Virtual Worlds* (MIT Press 1994).

93 Karl Popper, *Unended Quest: An Intellectual Autobiography* (Fontana 1976).

95 William H. Calvin, "A stone's throw and its launch window: Timing precision and its implications for language and hominid brains," *Journal of Theoretical Biology* 104: 121–135 (1983) at WilliamCalvin.com/1980s/1983JTheoretBiol. htm.

97 "Enlarge one neocortical area, enlarge them all" paraphrased from: Barbara L. Finlay and R. B. Darlington, "Linked regularities in the development and evolution of mammalian brains," *Science* 268: 1578–1584 (1995).

97 William H. Calvin, George A. Ojemann, *Conversations with Neil's Brain* (Addison-Wesley 1994).

9. FROM AFRICA TO EVERYWHERE

105 John E. Pfeiffer, *The Creative Explosion* (Harper and Row 1982), 207–208.

108 Creation myths: Melville J. Herskovits, *Man and his Works* (Knopf 1952), 68–69.

108 Timing for out of Africa: "These results indicate that male movement out of Africa first occurred around 47,000 years ago. The age of mutation 2, at around 40,000 years ago, represents an estimate of the time of the beginning of global expansion," Russell Thomson et al., "Recent common ancestry of human Y chromosomes: Evidence from DNA sequence data," *Proceedings of the National Academy of Sciences* (U.S.) 97(13): 7360–7365 (20 June 2000).

111 Fishing implements: Alison S. Brooks, et al., "Dating and context of three Middle Stone Age sites with bone points in the Upper Semliki Valley, Zaire," *Science*, 268: 548–553 (1995).

111 Beads: Stanley H. Ambrose, "Chronology of the Later Stone Age and Food Production in East Africa," *Journal of Archaeological Research* 25(4): 377–392 (1 April 1998). Cross-hatching on red ochre: Christopher S. Henshilwood et al., "Emergence of Modern Human Behavior: Middle Stone Age Engravings from South Africa," *Science* 295: 1278–1280 (2002).

111 Peter J. Richerson & Robert Boyd, "The Pleistocene and the origins of human culture: Built for speed," In *Perspectives in Ethology*, Volume 13. Nicholas S. Thompson and Francois Tonneau, eds. (Kluwer Academic/Plenum Publishers, New York. 2000), 1–45.

112 Spencer Wells et al., "The Eurasian Heartland: A continental perspective on Y-chromosome diversity," *Proceedings of the National Academy of Sciences* (U.S.) 98: 10244–10249 (2001). And see Spencer Wells, *The Journey of Man* (Princeton University Press 2003).

112 Stewart Brand, *Whole Earth Review* 68 (Fall 1990).

117 Elizabeth F. Loftus, "Creating false memories," Scientific American 277(3):

70–75 (September 1997) and her *Eyewitness Testimony* (Harvard University Press, revised edition 1996).

120   Fyodor Mikhailovich Dostoevsky, *Notes From Underground* [Letters from the Underworld] (1864). See kuyper.cs.pitt.edu/d/dostoevsky/underground/underground11.txt.

122   "Martyrdom is often the result of excessive gullibility." I have lifted this felicitous phrase from page 164 of *The Mummy Case*, a 1985 novel by the egyptologist Elizabeth Peters.

122   Robert Wright in *Slate* 9/2002.

123   Desmond Morris, *The Human Animal: A Personal View of the Human Species* (BBC Publications 1994).

123   Carl Sagan, *The Demon-haunted World* (Random House 1996), 26.

10. HOW CREATIVITY MANAGES THE MIXUPS

126   Joan Didion, *Michigan Quarterly Review* 18(4):521–534 (Autumn 1979).

128   Tattersall (1998), 28.

129   Language onset: Kenneth P. Oakley, *Man the Tool-maker* (University of Chicago Press 1949).

130   Marc D. Hauser, *Wild Minds* (Holt 2000), 62.

131   My six essentials (from *The Cerebral Code* 1996) build on the three that Alfred Russel Wallace listed in 1875 (". . . the known laws of variation, multiplication, and heredity . . . have probably sufficed. . . ."); I make explicit the pattern, the work space competition, and the environmental biases. See Wallace's "The limits of natural selection as applied to man," chapter 10 of *Contributions to the Theory of Natural Selection* (Macmillan 1875).

134   Mithen (1996).

135   Calvin (1996) discusses the cortical recruitment of a plainchant choir.

135   The technical term for this is coherence, as in a coherent light pipe, but here I have avoided my terminology from *The Cerebral Code* because here I am using coherence in another sense, of where everything hangs together in the output of the Darwin Machine.

137   Konner (2001), xviii

137   Steven Harnad, "Back to the Oral Tradition Through Skywriting at the Speed of Thought" (2003) at www.interdisciplines.org/defispublicationweb/papers/6

137   Galileo, *Dialogue concerning the two chief world systems* (1632, as reprinted University of California Press, Berkeley 1967).

137   Pinker (2002), 222.

11. CIVILIZING OURSELVES

140   Alexia without agraphia: William H. Calvin, George A. Ojemann, *Conversations with Neil's Brain* (Addison-Wesley 1994), 232–233.

142 George A. Ojemann, "Some brain mechanisms for reading," in *Brain and Reading*, edited by Curt von Euler (Macmillan 1989), 47–59.

145 Pinker (2002), 221.

146 Richard D. Alexander, *Darwinism and Human Affairs* (University of Washington Press 1979).

149 Konner (2001), 3.

149 David Brin, *Salon* (17 December 2002), at www.salon.com/ent/feature/2002/12/17/tolkien_brin/index.html

149 David Brin, *Tomorrow Happens* (NESFA Press, 2003).

12. WHAT'S SUDDEN ABOUT THE MIND'S BIG BANG?

150 John Maynard Smith and Eörs Szathmáry, *The Origins of Life: From the Birth of Life to the Origin of Language* (Oxford University Press 1999), 16–19.

152 Gossip, social life, and brain size: Dunbar (1996).

13. IMAGINING THE HOUSE OF CARDS

160 Pinker (2002), 70.

161 Proximate and ultimate causation, see Ernst Mayr, *The Growth of Biological Thought* (Harvard University Press 1982).

161 Levels of organization: Heinz R. Pagels, *The Dreams of Reason: The Computer and the Rise of the Sciences of Complexity* (Simon & Schuster 1988).

164 Daniel Chandler, *Semiotics for Beginners* at www.aber.ac.uk/media/Documents/s4b/sem07.html

165 Thomas Sprat, *The History of the Royal Society of London for the Improving of Natural Knowledge* (1667).

166 Dennett (2003), 266.

168 Turner (1996), 67.

168 Turner (1996), 114.

14. THE FUTURE OF THE AUGMENTED MIND

170 Joel Garreau (2003), at www.edge.org/q2003/. Tattersall (2002), 194.

176 William H. Calvin, *The River That Flows Uphill* (Macmillan 1986), 453–454.

176 Assortative mating is a common biological phenomenon. But whether there is actual IQ shift or not also depends on how many children those high-high couples have, compared to otherwise.

177 Ernst Mayr, *The Growth of Biological Thought* (Harvard University Press 1982), 47.

177 Erich Jantsch, *Self-Organizing Universe: Scientific and Human Implications* (Pergamon Press 1980).

178 Elizabeth Peters, *He Shall Thunder in the Sky* (Morrow 2000), 200.

179 Russell Gardner, Jr., "Evolutionary perspectives on stress and affective disorder," *Seminars in Clinical Neuropsychiatry* 6(1): 32–42 (2001).

179   Randolph M. Neese, "Is depression an adaptation?" *Archives of General Psychiatry* 57: 14–20 (2000).

185   Adam Zeman, *Consciousness: A User's Guide* (Yale University Press 2003), 181.

185   William H. Calvin, "Competing for consciousness: A Darwinian mechanism at an appropriate level of explanation," *Journal of Consciousness Studies* 5(4)389–404 (1998).

187   Connie A. Woodhouse, Jonathan T. Overpeck, "2000 years of drought variability in the central United States," *Bulletin of the American Meteorological Society* 79(12)2693–2714 (1998).

187   Barbara Kingsolver, *Small Wonder* (HarperCollins 2002), 101. And see Jared Diamond's *Ecocide* (Allen Lane 2004).

190   Stewart Brand's statement of purpose for the *Whole Earth Catalog* (that won the National Book Award in 1972) was, as usual, more succinct: "We are as gods and might as well get good at it."

Index

habitat, 16, 18, 23
Hadar (Ethiopia), 24
Hadza people of Tanzania, 35, 200
hair, 18
hallucinations, 179
hammering, 47, 54, 58, 64, 66, 75, 98
handaxe, 41–43, 199
hands, 20
hardwiring, 142
harmony, 93, 94, 142
Harnad, Steven, 137, 191, 205
Hauser, Marc, 21, 51, 86, 130, 195, 199,
    200, 203, 205
Hawking, Stephen, xiv, 197
headlights, 115, 174, 175, 181, 186
heating, 18
hidden order, 90
hidden patterns, 93, 114
historians, xiv, 139
hogtie, 115
hominid, xix, 198, 199, 201, 202, 203
hominin, xix
Homo erectus, xx, 16, 17, 19, 20, 33, 34,
    36, 37, 38, 40, 41, 43, 45, 53, 56, 57, 63,
    64, 110, 112, 199
Homo ergaster, 36
Homo habilis, 35
Homo heidelbergensis, 53, 56
Homo rudolfensis, 35
Homo sapiens (anatomically modern)
    xiii, 55, 62, 67, 81, 83, 199
Homo sapiens sapiens (behaviorally
    modern as well), xiii, xiv, 60, 62, 90,
    107, 110, 112, 116, 157, 171, 190, 197
honor, 39
Hopkins, William, 191
horizontal transmission, 158
Hrdy, Sarah, 191, 195
humor, 164, 166
Humphrey, Nick, 194
hunter, 34, 39, 43, 66, 68, 159, 202
hunting, 7, 10, 12, 17, 26, 30, 33–42, 47,
    49, 50, 55, 67, 70, 200–202

hyena, 22
hypoglossal nerve, 56
hypotheses, 78

ice, xiii, 23, 40, 46, 47, 61, 63, 83, 123, 178,
    199, 202
ice age, 40, 46, 47, 61, 63, 83, 157, 199, 202
imagination, 108, 169, 186
imitation, 8, 71, 72, 73, 85, 202
immigrants, 27, 29, 132
inbreeding, 133
incoherence. *See* coherence.
Indo-European languages, 109
infants, 8, 18, 116, 203
infection, 102, 156, 157
inference, 91, 99
inflections, 85, 88
information, sharing, 158
ingestion, 28
injuries, 55
innovation, 8, 28, 29, 52, 90, 103, 124,
    127, 128, 136, 158, 174
instabilities, 181
instinct, 90
instincts, 9, 30, 90, 114, 140–142, 166,
    168, 178, 179
insulation, 26, 57
intellectual function, xv, 10, 77, 83, 94,
    103, 104, 114, 116, 127, 134, 144, 150,
    157, 166, 179, 185, 186
intelligence, xvi, 9, 34, 38, 53, 68, 70, 79,
    94, 99, 108, 123, 127, 135, 136, 166, 177,
    181, 184, 202. *See also* IQ.
intelligence, artificial, 143
intelligence, Machiavellian, 135
interbreeding, 111
interconnections, 135, 156
intermediate product, 40
intersubjectivity, 76
intuitions, 52, 143–145, 167, 168
intuitive notion of probability, 144
intuitive physics, 97, 143
intuitive psychology, 144